ASSET MANAGEMENT IN THE SOCIAL RENTED SECTOR

Asset Management in the Social Rented Sector

Policy and Practice in Europe and Australia

Edited by

Vincent Gruis
Faculty of Architecture, Department of Real Estate & Housing,
Delft University of Technology,
The Netherlands

and

Nico Nieboer
OTB Research Institute for Housing,
Urban and Mobility Studies,
Delft University of Technology,
The Netherlands

KLUWER ACADEMIC PUBLISHERS
DORDRECHT / BOSTON / LONDON

A C.I.P Catalogue record for this book is available from the Library of Congress

ISBN 1-4020-2557-2 (HB)
ISBN 1-4020-2558-0 (e-book)

Published by Kluwer Academic Publishers,
P.O. Box 17, 3300 AA Dordrecht, The Netherlands.

Sold and distributed in North, Central and South America
by Kluwer Academic Publishers,
101 Philip Drive, Norwell, MA 02061, U.S.A.

In all other countries, sold and distributed
by Kluwer Academic Publishers,
P.O. Box 322, 3300 AH Dordrecht, The Netherlands.

Cover picture:
Amsterdam, Buitenveldert, by Axel Smits, OTB Research Institute

Printed on acid-free paper

Printed in the Netherlands.

CONTENTS

PREFACE

The idea to start an international investigation into asset management in the social rented sector was triggered by developments that we observed in our own country—the Netherlands. During the nineties, the Dutch government reduced its regulation as well as its financial support of social landlords. Furthermore, the market share of social rented housing was under threat due to the booming economy and a growing preference for owner-occupation. These developments lead to a widespread interest among Dutch social landlords in the adaptation of more professional, business-like approaches towards the management of their housing stock. Since similar developments had taken place in other countries as well, we felt it would be interesting to see if these developments had a similar impact on the social landlords, and if there are opportunities for international exchange of experiences. However, a literature review quickly revealed us that little had been published about this topic—most comparative housing research being focused on national housing policies and systems, often placed in the wider context of the analysis of welfare regimes. So we had two options. The first was to conduct an investigation into the asset management in various countries ourselves. This would have meant a lot of nice travels, but also a lot of time, money and the risk of not being able to paint a reliable picture because of our unawareness of specific characteristics of the social housing system, policy and landlords in other countries. Therefore we chose the second option, which was to gather a group of people who would be willing to write about their own country in answer to our research questions. To meet potential contributors we attended the conferences of the European Network for Housing Research (ENHR) in 2001 and 2002. At the latter conference we organised a workshop on asset management in the social rented sector during which we met the majority of the authors of this book. At the ENHR conference in 2003 we held a focused follow-up workshop to discuss the draft chapters and now, in 2004, the final result lies before you. So, there have been four years, three conferences and uncountable emails between the initial idea and the resulting book, but we feel it has certainly been worth while. The book fills a gap in international comparative housing research, gives means to place one's own national policy and practice in a wider context, is a useful starting point for a more structural international exchange of experiences and is a basis for further theoretical and practical development of asset management in the social rented sector.

This project is a part of Delft University of Technology's research programme on Sustainable Urban Areas and has been conducted within the ENHR Working Group on Housing Management. Of course, the book would not have been possible without the contributions from all the authors—many thanks for your cooperation! Furthermore, we are very grateful to the Dutch Ministry of Housing, Spatial Planning and the Environment for its financial support. Finally we would like to thank our colleagues within our university for their support of the book.

Vincent Gruis and Nico Nieboer
May 2004

1 INTRODUCTION

VINCENT GRUIS and NICO NIEBOER

1.1 SCOPE AND AIMS OF THE BOOK

Social housing management is changing in response to developments in its institutional and economic context. In many countries, housing systems are in transition as part of a more general trend towards privatisation and decentralisation of public services. These transformations have lead to a more market-oriented social housing management (e.g. Priemus et al., 1999). In Britain, for example, housing associations gained the governments preference as social housing providers above municipalities, among others because they are more detached from the government and are more like 'real' market parties (Walker and Van der Zon, 2000). In Germany, the distinction between social and commercial landlords has faded with the act that repealed the Public-Use Housing Law ("Wohnungsgemeinnützigkeitsgesetz") in 1984 (e.g. Kleinman, 1996). In Australia the current Commonwealth-State Housing Agreement allows State housing authorities greater flexibility in determining housing strategies which are appropriate for their State, as opposed to the attempt to impose national consistency which had characterised previous arrangements (e.g. Larkin, 2000, p. 46). These reforms in the housing systems have often been accompanied by a reduction of government finance and subsidisation as we can see in for example in Sweden, England, The Netherlands and Australia (Boelhouwer, 1997, 1999; Smith and Oxley, 1997; Larkin, 2000). As a result, social landlords have become more dependent of the private capital market and are forced to operate in a more business-like way to keep their financiers satisfied (e.g. Walker and Van der Zon, 2000; Boelhouwer, 1999).

In parallel with developments in public policy, there has been a general trend towards a decreasing market share for social rented housing, partly due to economic growth and a shift towards owner-occupation. Furthermore, many housing markets are changing from one of general shortage to one in which over-supply occurs in certain local sub-markets But regional differences are great; there are still substantial social housing needs that social landlords have to fulfil (Priemus et al., 1999). Furthermore, large parts of the European housing stock are encountering problems. As, for example, Thomsen and van der Flier (2002, p. 2–3) point out: "The majority of the housing stock in the European Union was built after the second world war, varying from 52% in

Vincent Gruis and Nico Nieboer (eds.),
Asset Management in the Social Rented Sector, 1–18.

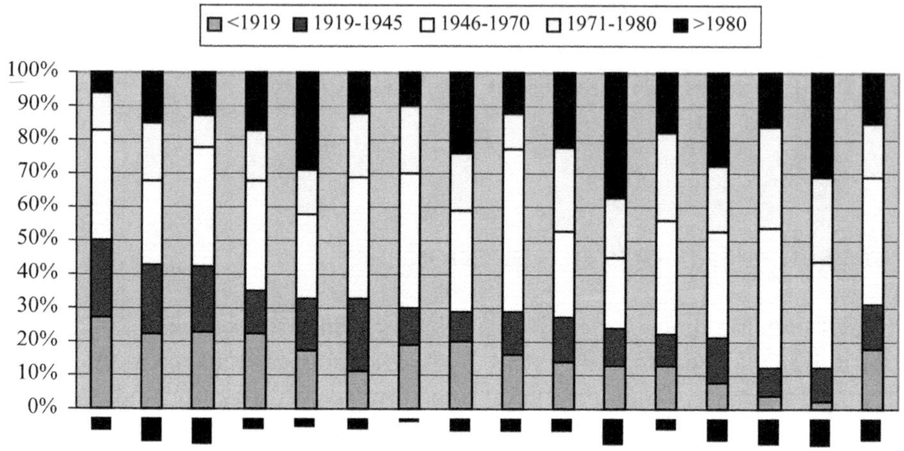

Figure 1.1 Composition of the EU housing stock by building year (1998)
Source: Haffner (1998); data processed by Thomsen and van der Flier (2002).

the UK to 75% in the Netherlands and 85% in Finland (Figure 1.1). After WW-II, most EU countries have countered housing shortages by large scale housing production of modest quality, mainly apartment-blocks for 'standard' family housing. This massive housing production has caused a massive ageing process, resulting in physical and, above all, functional decay: too many small, uniform and poorly equipped flats in dull apartment blocks ... ". For social landlords these changes in segments of the housing market mean that they have to anticipate future levels of demand to prevent vacancies and to adjust to new markets.

The changes in the institutional and economic context set considerable challenges for the asset management of social landlords. In the past, social housing was mainly concerned with the development of new dwellings. Management of the existing stock consisted first of all of day-to-day maintenance and administration. Strategic decisions about investments in the housing stock were made mostly by governments, rather than by individual landlords. Now, the emphasis is shifting towards management of the existing stock—new housing production has become a marginal phenomenon when set against the size of the housing stock (see Figure 1.2)—and landlords throughout Europe and Australia are faced with the challenge to develop their own strategies, although in some countries more prominent than others. However, being a relatively new topic in the social rented sector, international comparative research about the asset management of social landlords is scarce. International literature has been concerned mainly with the description of the changes in the political and economic context of social housing management and the 'privatisation'

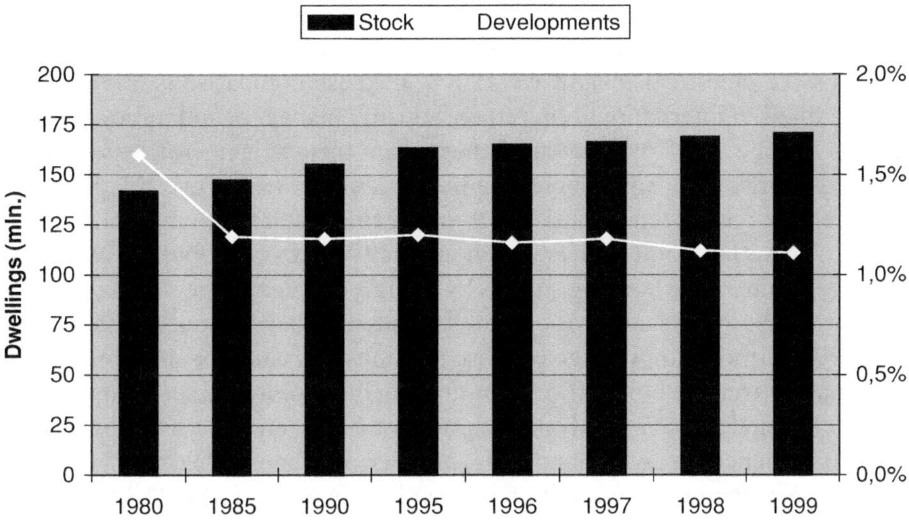

Figure 1.2 Annual new constructions as percentage of the existing EU housing stock
Source: Haffner (1998); data processed by Thomsen and van der Flier (2002).

of social housing. Some reflections have been made on the general consequences of these developments for social landlords (e.g. Boelhouwer et al., 1997; Priemus et al. (eds.), 1999), but a descriptive or a comparative study of the asset management practice of social landlords has not been conducted.

The main objective of this book is to fill the gap in international literature. The book aims to provide insight in asset management in the social rented sector, by describing and comparing the practice of social landlords[1] in various (mainly European) countries. The book focuses on the strategic aspects of asset management: the strategy development at the top-level of the landlords' organisation. It describes and compares how social landlords develop strategies for the management of their housing stock, by looking at the process followed, the strategies formulated and the objects, means, methods, instruments and criteria used in different aspects of asset management (rent setting, allocations, sales, maintenance and renewal). The practice of social landlords is

[1] This book is concerned with the management of the social rented housing stock. We point out that the social rented stock can be different from the stock of social landlords. Some parts of the social rented stock are entirely managed by government institutions (such as the local authorities in England and the State housing authorities in Australia), other parts by private landlords (as is the case in Germany) or 'semi-public' organisations (such as the Dutch housing associations). Thus, we use a broad definition of social landlords: namely all private, public and semi-public institutions that are responsible for the management of social rented housing stock.

set within their national (political, institutional and economic) context. These comparisons are interesting from a theoretical and a practical point of view. Besides the empirical insight we gain with these comparisons and the gap that is filled in international literature, housing managers and policy-makers across Europe and Australia can benefit from this exchange of experiences. The comparison between social landlords can also (dis)prove relationships between their approach towards asset management and their national context. From a scientific point of view, such an analysis can contribute to the more general comparative housing studies by gaining a deeper understanding of the way in which social landlords operate is related to their national housing system. Furthermore, knowledge of these relationships could be used to support national housing policies: if certain approaches towards asset management are recognised as having advantages over others, then governments may be able to influence the social landlords' practice by creating the 'right' context. In summary, the main research questions that are addressed in this book are:

- How do social landlords in different countries formulate strategies for the management of their housing stock?
- What are the relations between the landlords' practice and their national context?

The book consists of three parts: 1) a theoretical framework (in this introduction), 2) a number of country monographs about asset management practice of social landlords and their national context and 3) comparative conclusions.

In the following sections we describe our theoretical framework. First we define asset management in a social housing context. Based on a general model for business planning, we describe the activities that can be carried out to formulate asset management strategies. Then, we set up a typology of possible approaches towards asset management and we construct a hypothetical model about the relations between the landlords' practice and their national housing context, which we will use as an analytical framework for our comparative conclusions.

1.2 ASSET MANAGEMENT IN THE SOCIAL RENTED SECTOR: DEFINITION AND ACTIVITIES

Priemus et al. (1999, p. 211) define social housing management as "the set of all activities to produce and allocate housing services from the existing social housing stock". According to Priemus et al., housing management consists of a variety of activities, categorised in technical management (maintenance, renovation, etc.), social management (housing allocation, etc.), financial management (treasury, rent policy) and tenure management (letting, buying, selling) (p. 212). We follow this definition but refer to 'asset management in the social rented sector' instead of (the shorter term) 'social housing management' since the latter is an ill-defined concept. While social housing management is unclear in its subject (e.g. Scott et al., 2000; Pearl, 1997), asset management seems to be clear in its focus on the physical housing stock. Asset management includes activities in all the four categories mentioned by Priemus et al., but excludes activities that do not affect the characteristics of the housing stock (e.g. treasury management and business administration). The main activities in asset management concern maintenance, renewal, sale, allocation and rent policy.

Asset management is a relatively new concept in social housing. It stems from the private sector where it is concerned with an analysis of the performance of an organisation's assets in support of decisions about holding, selling and repositioning. In private sector asset management, the emphasis is on optimising financial performance. In the social rented sector, financial performance is not necessarily—or mostly not—the primary criteria for management decisions (although the current developments imply it has gained importance among social landlords in various countries). The key-question for social landlords is how to reach their social housing objectives efficiently (Gruis and Thomas, 2002). Hence, it is also interesting to note Larkin's (2000, p. 8) definition of asset management in the context of social housing as "the range of activities undertaken to ensure that the housing stock meets needs and standards now and in the future in the most efficient way".

Priemus et al. (1999) also distinguish day-to-day (operational) management and strategic housing management. The latter concerns medium and long-term management policies, usually formulated at a strategic (top-management) level in the organisation. This distinction can be clarified by making a comparison with the distinction made within (commercial) real estate management between portfolio, asset and property management (see Figure 1.3).

Within Figure 1.3, Miles et al. (1996) place 'portfolio management' at the top, strategic level of the triangle. However, this term is used by (commercial) investors to determine the allocation of their resources to several investment categories (shares, bonds, real estate, etc.) at a meta-level within the organisation and is thus not really applicable to social landlords, if only because they are by definition restricted to investments in one specific category of real estate: housing. Hence, in analogy with Priemus et al. (1999), but focussing on the housing stock, we use the term 'strategic asset management'. Within Figure 1.3, strategic asset management may be placed best on the level of asset management, but it also contains elements of portfolio management as

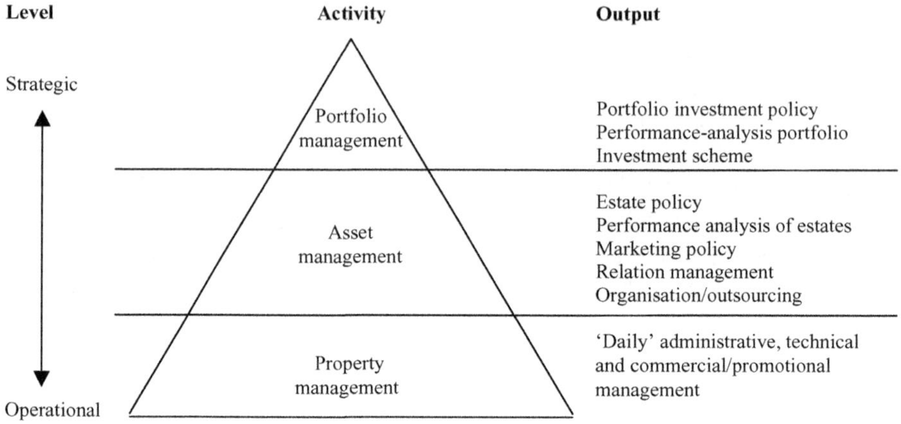

Figure 1.3 Organisational levels of real estate management
Source: Miles et al. (1996).

far as the residential portfolio is concerned: it is concerned with defining the desired mix of housing (dwelling types and prices), analysing the performance of the residential portfolio, defining guidelines for management, acquisition and disposition of the estates in the portfolio.

1.3 A TYPOLOGY OF APPROACHES TOWARDS ASSET MANAGEMENT

The distinction of strategic asset management is the starting point for our typology of approaches towards asset management. The term 'strategic asset management' has been introduced in the Dutch social rented sector in the early 1990s. Since then, as a result of housing reforms, social landlords in the Netherlands have gained a considerable degree of administrative and financial independence. In addition, the market position of the social rented sector was weakening in the 1990s due to the booming economy and a shift in housing preferences towards owner-occupation. These developments lead to the introduction of principles from *strategic planning* in support of the management of the social rented stock.

Strategic planning is the process of developing and maintaining a viable fit between the organisation's objectives and its recourses (Hannagan, 1992, p. 38). Although it has its origins in the private sector, this concept has also been introduced in the management of public organisations (see e.g. Bryson, 1995). Many advantages are ascribed to following a strategic approach towards business planning. As Fraser and Stupak (2002, p. 1203) put it, "advocates of strategic planning believe the process will amplify and enhance systematic information gathering, clarification of organisational direction, establishment of priorities, quality decision making, communication and understanding of strategic intent, solid organisational responsiveness, effective performance, conscientious framework, useful application of expertise, and attention to organisational learning." In short, strategic planning is expected to contribute to an organisation's effectiveness and efficiency by following a systematic, rational and transparent planning process.

Strategic asset management combines principles of (commercial) asset management and strategic planning to develop methods and tools for management of the social housing stock. Strategic asset management can be typified further on the basis of interrelated characteristics that can be found in literature on strategic (business) planning. These business planning characteristics form the basis of our evaluation framework and can be summarised as market oriented, systematic, comprehensive and proactive. Below we reflect on these characteristics and their potential benefits for social landlords (see also Gruis et al. 2004):

Market-oriented

Asset management in the social rented sector is concerned with fulfilling a housing demand by offering a housing supply and in that sense it is market-oriented by definition. However, social housing has traditionally been provided through 'bureaucratic' mechanisms (government regulation and subsidies) and not by market forces. In general, literature on strategic planning emphasises the need of an analysis of a companies' own strengths, weaknesses in relation to the opportunities and threats in their (market) environment in support of strategy formulation (e.g. Aaker, 1998; Kotler, 1997; Bryson, 1995). In the private sector, strategies are based on an analysis of the market position of the products, market prospects and—in general—opportunities to earn money. In the 'classic' portfolio analysis, designed by the Boston Consultancy Group, cash-performance is crucial in the analysis of business units (see Ansoff, 1984). In analogy with commercial practice, a market-oriented landlord can be expected to place more emphasis on analysing market demand and opportunities. Important decision-making factors in strategy formulation will be current lettability, future market expectations, financial return and opportunities for sale. A wide range of strategies will be considered and applied: diversification of the price and quality of dwellings within the portfolio according to housing demand will be a central theme in asset management.

A market-orientated landlord can be set against a task-oriented or responsive landlord whose focus is mainly on fulfilling 'traditional' social housing tasks: the letting of decent, affordable dwellings. This distinction between task-oriented and market-oriented is somewhat comparable to the distinction made by Miles and Snow between 'Defenders' and 'Prospectors', which they characterise as follows: "*Defenders* are organisations which have a narrow product-market domains. Top managers in this type of organisation are highly expert in their organisations limited area of operations but do not tend to search outside of their domains for new opportunities. As a result of this narrow focus, these organisations seldom need to make major adjustments in their technology, structure, or methods of operation. Instead they devote primary attention to improving efficiency of their existing operations. *Prospectors* are organisations which almost continually search for market opportunities, and they regularly experiment with potential responses to emerging environmental trends. Thus, these organisations often are the creators of change and uncertainty to which their competitors must respond." (Miles and Snow, 1978, p. 29). Of course, it should be acknowledged that the specific characteristics of social landlords and (government) regulation of the social rented sector do not allow social landlords to behave like commercial enterprises. They are for example restricted to offering (social) housing and (hence) financial return is not their primary objective. Nevertheless, within these boundaries, increased market orientation can have benefits for social landlords as well. Market orientation can help social landlords to realise a portfolio which is effective (in meeting housing demand and tenants' preferences) and economically efficient (using 'cash cows' to finance the core social housing stock).

It is also interesting to compare the concept of market orientation with Kemeny's discussion of the concept of market-conformity in the context of the social market theory. This theory has been developed in the 1930s in Germany "as an alternative to the extremes of classical liberalism and the command economy (Willgerodt and Peacock, 1989) (...) [Its] proponents wished to develop a 'third way' (...) [which] involved constructing markets in such a manner as to ensure that important social goals were built into the market (...) based on the principle that intervention is necessary and desirable, but that it must be *marktkonform* or market-conforming (Müller-Armack, 1989, p. 84)" (Kemeny, 1995, p. 11–12). Kemeny acknowledges that the concept of market-conformity is difficult to specify theoretically and illustrates this concept with examples of market-(in)conform behaviour. Therefore, and certainly in the context of asset management, we argue it is better to use the term 'market-oriented' instead of 'market-conform' to avoid potential mistakes with outright commercial, profit-seeking behaviour. The latter implies that social landlords would strive to optimise their financial return and act as private investors and, consequently, would not invest in social housing if they were not sufficiently compensated by subsidies. Market-oriented is much 'softer' in a sense that social landlords try to contribute to solving shortages and surpluses on the housing market, including profitable housing investments, but emphasising social housing needs.

The occurrence of market orientation can be reflected in the various activities of social landlords' asset management: rent(increases) will be related to the quality and market position; allocations, maintenance and renewal activities will take market demand and tenants' preferences into account and landlords will have an active sale policy to generate financial income and meet housing preferences.

Systematic

Many books on strategic business planning suggest the use of systematic planning procedures and rational frameworks for decision-making. Thus, within strategic asset management, a landlord will put effort into rational and transparent decision-making. The process of formulating asset management strategies will be well-structured. Decision-making factors will be clearly marked and the way in which decisions are reached will be reported. Asset management decisions have a large influence on the quality, affordability and availability of dwellings, being the key-objectives of social housing everywhere. This impact on social housing objectives places a demand on the quality of the decision-making process. Stakeholders of social landlords may expect a 'justifiable' policy, which is supported by rational arguments as part of a transparent decision-making process. In fact, it can be argued that social landlords should strive towards such a transparent policy as part of their social objectives. The occurrence of a systematic approach towards asset management can be reflected in the application of decision-making frameworks–comparable, for example, with private sector portfolio analyses—and structured processes—comparable for example with strategic business planning as described by Kotler (1997, p. 80), Aaker (1998, p. 19) and Bryson (1995, pp. 22–37).

Comprehensive

A major characteristic of strategic business planning is that it deals with the objectives of the organisation as a whole, at top-management level. In general, models for strategic planning include the formulation of a mission statement and business goals to guide the development and activities of an organisation. Products or 'business units' are analysed and compared with each other in the light of this mission. As stated at the beginning of this section, asset management concerns only a part of a social landlords' activities. Nevertheless, the characteristic of 'comprehensiveness' can be applied to the specific area of asset management. Comprehensive asset management will focus not only on individual dwellings or estates, but will also reflect on the composition of the stock as a whole. Furthermore, different aspects of stock management will be considered. For example, technical and social activities, long-term and short-term objectives, and activities at a strategic and operational level. A landlord operating in only a responsive way will focus for example on problem estates, failing to formulate objectives for the development of the whole housing stock and will not consider the (lack of) synthesis of different parts of the total management approach. A comprehensive (portfolio) approach helps social landlords to determine which part of the stock should be given priority for investment and intensive management. Furthermore, reflection on the desired growth direction of the portfolio as a whole, in relation to housing needs, allows them to put decisions about individual estates in a wider perspective.

Proactive

The final key-characteristic of strategic asset management that we consider in our evaluation framework is the occurrence of a pro-active approach. Books on strategic planning are very clear on the fact that strategic behaviour is not about taking a passive attitude towards developments in the company's environment. For example Ansoff (1984, p. xv) states strategic planning "is a systematic procedure for management which anticipates the challenge and prepares its responses in advance, based on examination of novel alternatives". Strategic planning can be set between 'long-range planning', which is based on (sometimes inadequate) extrapolative forecasts, and 'strategic management', which suggests real-time strategic response to the dynamic environment (e.g. Ansoff, 1984, Aaker, 1998). It is interesting to note that, from a historic perspective, strategic management has evolved from strategic planning, because the 'planning cycle' used in strategic planning is assumed to be "inadequate to deal with the rapid rate of change that can occur in a firm's external environment" (Aaker, 1998, p. 11). Although the inflexible nature of the housing stock can impede a quick response to changes in the environment, we recognise pro-activeness as a key-characteristic of both strategic planning and management. Translated to asset management in the social rented sector, a proactive approach implies that landlords will actively identify problems and opportunities stemming from developments in the housing market, housing policy and market position. Furthermore, they will anticipate these developments with asset management strategies instead of reacting to them after potential problems have become reality (for example, initiating renewal before a neighbourhood begins to deteriorate).

In summary, we use the term 'strategic' for landlords who follow a market-oriented, systematic, comprehensive and proactive approach towards asset management. In Table 1.1, we have summarised possible (qualitative) 'indicators' of the occurrence of these characteristics. The four characteristics and their opponents provide the basis for our typology of approaches towards asset

Table 1.1 Characteristics and 'indicators' of strategic asset management

Characteristic of strategic asset management	'Indicator' of occurrence
Market-oriented	Rents, allocations, sales, maintenance and renewal are related to tenants' preferences, market demand and financial return/opportunities.
Systematic	Frameworks for decision-making and (structured) planning processes are applied.
Comprehensive	Goals are formulated for the development of the entire housing stock and individual estates are analysed in relation to each other.
Proactive	Investments and other activities anticipate threats and opportunities.

Source: Gruis et al. (2003).

management, which is shown in Figure 1.4. In the conclusion of this book we will use this typology to reflect on the current practice of social landlords in the investigated countries. Of course, in practice mixed forms exist with some landlords, for example, acting in a comprehensive and systematic way but not in a market-oriented manner. Hence it is often not possible to classify landlords rigidly into 'strategic' and 'non-strategic', but the identification of the key-characteristics of strategic asset management from theory enables us to specify the subject of our comparison. Furthermore, it must be noted that our typology is rather prescriptive. Based on existing theories, and given the current interest in a more strategic asset management within the social rented sector in several countries, we have stressed the potential benefits of following a strategic approach towards asset management. The strategic planning approach is, however, not free of criticism (see e.g. Mintzberg, 1994—compare with the discussion on rational and incremental approaches). Furthermore, application of some of the characteristics in the social rented sector could 'backfire'. This counts in particular for market orientation: if this leads

Figure 1.4 Characteristics of strategic asset management and their 'opponents'

to market conformity in a sense that social housing tasks are forgotten and profit-making becomes the leading motivation, social landlords will loose their position as effective contributors to the fulfilment of social housing needs. We will return to the potential pitfalls of strategic asset management in the social rented sector in our conclusion (section 11.4).

1.4 LANDLORDS' PRACTICE AND THEIR NATIONAL CONTEXT: A HYPOTHETICAL MODEL

The typology of approaches to asset management provides a touchstone for comparison of practice of different landlords in different countries. This comparison becomes particularly interesting if we relate them to characteristics of the countries' housing systems. There can be, for example, a relationship between the measure of government regulation and the kind of asset management: landlords in countries that allow them much more freedom in determining their own policies can be expected to operate in a more 'strategic' way (even if only to compensate for the lack of government regulation). In this section we discuss a hypothetical model to identify possible relations between the national context and landlords' approaches to asset management. Figure 1.5 contains a schematic representation of our model; here follows an explanation.

In describing the relevant factors of the national context we distinguish two main variables: the housing system and the housing market. 'Housing system' is used as a general term for the whole of public housing policy, regulations, legislation, financial government support and institutional arrangements, including the historic developments within the national system. These variables

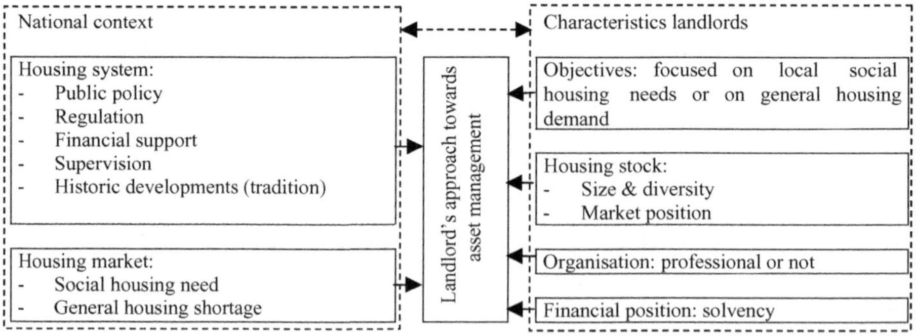

Figure 1.5 Hypothetical relations between landlords' characteristics, national context and approach towards asset management

can influence the way in which landlords operate in various, interdependent ways:

- Strong government regulation (through grants, legislation, renewal programmes, land policies etc.) means less opportunity and necessity for landlords to make their own policy;
- Substantial financial government support and guarantees (in addition to its function in government regulation), gives landlords little incentive for strategic behaviour in which they avoid (financial) risks and seek opportunities for increasing their financial return or economic efficiency.
- Supervision and institutional arrangements can also have consequences for (the lack of) strategic behaviour of landlords. Social landlords who operate within an institutional structure in which strategic decisions are made by the government or central housing organisations and are solely responsible for carrying out daily management of the (marginal) social housing stock have little need for strategic asset management.
- Historic developments of the housing system must be viewed as well. If landlords have only recently gained financial and regulatory independence, it can be expected that their organisations have yet to adapt to the new situation. Substantial differences between front- and back-runners can be expected.
- Landlords that operate within a more relaxed housing market have to deal with increasing management problems. Strategic asset management will be necessary to prevent and cure problem estates.

Next to their context, characteristics of (individual) landlords can also influence their approach towards asset management:

- Landlords' objectives (as part of regulatory and institutional arrangements as well as their own business mission) may be reflected in their approach. Landlords who have a broad function on the housing market, supplying housing services for a diverse group of households, can be expected to operate more strategically to cope with this diversity (which will also be reflected in the diversity of the stock).
- Landlords who operate a stock with a poor quality, increasing social problems and/or a threatened market position will need strategic management to cope with these problems.
- Landlords who have a professional management can be expected to operate more strategically than landlords whose management consists of volunteers and tenants.
- Landlords who operate under difficult financial conditions (as reflected by their solvency and financial government support) need strategic decisions

to be able to keep fulfilling their role as providers of decent, affordable housing.

Within our hypothetical model in Figure 1.5 we try to cope with the full complexity of (possible) relations between landlords' characteristics, housing context and approach towards asset management. However, it must be stressed that a full testing of this model is not feasible (at this moment) for two main reasons:

- The historic context in many countries makes it difficult to make conclusive statements at this moment. To date, in almost all of the countries that are involved in our research project the emphasis has been on the development of new social housing instead of management of the existing stock. Even in supposed front-runners such as the Netherlands, asset management is still in a development stage (Nieboer and Gruis, 2001). Organisations need a lot of time to adapt to their new situation. Consequently, the results of the country analyses must be captured in terms of 'tendencies towards strategic asset management' rather than in conclusive statements about the state-of -the-art.
- A full testing of the model implies a massive, in-depth investigation into the practice of many landlords within the countries involved in our research project, which simply is not feasible. Our analyses will be based on country monographs, written by national housing researchers, which will paint a general picture of landlords' practice, illustrated by a few case study examples.

Being aware of these limitations, it is not the objective of our research to (dis)prove all of the supposed relationships. Rather, our research is aimed at refining the hypothetical model by identifying key-factors which influence the social landlords' approach towards asset management.

1.5 CLASSIFICATIONS OF HOUSING SYSTEMS AND ASSET MANAGEMENT

Within comparative housing research, much attention has been given to the classification of national housing systems. Given that landlords' objectives, housing stock, organisation and financial position are not accidentally grown, but that they are part of the housing system in which they operate and given the objective of our research (looking for relations between landlords context and practice), it is useful to examine if we can find relationships between the type of housing system and asset management in the social rented sector. In this section we discuss if existing classifications of housing systems could be used to identify relationships between the type of housing system and the type of asset management. In doing so, we focus on the usefulness of two relatively well-known classifications of housing systems for this purpose, namely that of Kemeny and that of Harloe.[2]

Kemeny (1995, 2001) distinguishes two general types of (social) rental housing systems, which he describes as follows. The distinction stems from "two contrasting underlying philosophies in the state's role in housing provision in modern welfare states. In one philosophy, the state takes upon itself the direct responsibility of providing rental housing for households in need. To this end, non-profit rental housing is organised in the form of a state or local government monopoly. As far as possible, the non-profit sector is prevented from competing with private profit-seeking housing companies by hiving it off from the market into a command-economy public rental-housing sector. Access is in terms of 'need'—the definition of which varies over time and also between countries. The result is a 'dualist rental system' in which two distinct rental tenures are created, a 'public' command-economy sector and a 'private', and largely unregulated, rental market. In the other philosophy, the state is either not a major provider itself or, if it is, access to such housing—often provided on a 'not-for-profit' basis—is not limited to households in need. Instead it is encouraged to compete with profit-rental housing on the open market for tenants and thereby set standards, ensure that all households have security of tenure and competitively hold rents down. The result is that the legal distinction between profit and non-profit is minimised and a 'unitary rental market' is encouraged to emerge. But again, like the 'public housing only for those in need' philosophy, wide variations are found in the extent of the competition and the degree of non-profit influence" (Kemeny, 2001, p. 66; for further discussion see Kemeny, 1995).

Harloe (1995) distinguishes the mass model and the residual model in social housing provision. "The distinguishing feature of the mass model is its emphasis on a large volume of new construction of social rented dwellings, which are not specifically aimed at the lowest income households. Moreover, the sector is supported by generic subsidies; income dependent subsidies play a relatively insignificant role. The residual model, in contrast is characterised by small-scale programmes of new construction geared to the needs of the lowest-income groups. Harloe's residual model and Kemeny's dualist model have strong similarities with respect to their 'visible' effects" (Van der Heijden, 2002, pp. 328–329). However, while Harloe sees a convergence in the direction of the residual model in Western Europe, that has started in the mid 1970s,

[2] We are also aware of other, broader classification systems such as Esping-Andersen (1990) who introduces the concept of welfare regimes. Although we recognise the importance of this type of work, we have not used it because of the difficulties in applying this type of framework even to housing systems in general (see e.g. Hoekstra and Reitsma, 2002).

Kemeny believes a second model may develop (or remain), which can be roughly typified by his unitary model (Van der Heijden, 2002).[3]

For our research purposes, Kemeny's typology seems more useful than that of Harloe. While Harloe's typology is based strongly upon developments in the new construction of social housing over time, Kemeny provides a typology that can be applied to housing systems as they are at this moment. Moreover, since it is not only focused on construction, it can be applied to management as well. If we follow Kemeny's description of rental housing systems and underlying philosophies, *we can expect more strategic behaviour among landlords that operate within a unitary rental market.* Public landlords in a dualist rental system are characterised to a large extend by strong government regulation, financial support (or dependence). Thus, strategic decisions are taken by the government, rather than by the landlords. Furthermore, their objectives and housing stock are a more focused on the lower-income households, so there is little need for strategic management of a diverse portfolio. According to Kemeny, non-profit (or not-for-profit) landlords that operate within a unitary rental market are in fact encouraged to adopt market-oriented behaviour. The competition with private (profit) landlords, combined with relatively less government regulation leads to a greater need for a market-oriented management by the social landlords.

1.6 METHODOLOGY AND OUTLINE OF THE BOOK

In the conclusion of this book we analyse the asset management practice of social landlords in various countries in relation to their national context and characteristics. Our analysis will be based on the country monographs included in this book. The book focuses mainly on (North-) Western European countries, given the fact that the developments affecting housing management that are described in this introduction are (most) prominent in this part of Europe (see also Priemus et al., 1999; Boelhouwer, 1999). Cases that are included in this book are: Austria[4], Belgium, Denmark, England, France, Germany, and the Netherlands. Furthermore, we have included two 'contra-cases' for different reasons. We have included the Eastern European case of Latvia to illustrate the practice and specific challenges in former communist

[3] Another scenario is posed by Van der Heijden (2002) as the 'residual unitary market scenario' that starts with Kemeny's unitary market model, but develops in the direction of the residual or dualist rental market (which makes this scenario difficult to distinguish from Harloe's convergence theory).

[4] Although the geographical location of Austria may lend itself better for a qualification as a Central European country, we have included Austria among the Western European countries because of its traditionally 'western' regime (in contrast to Latvia).

regimes, referring in particular to the impact of mass-privatisation of the public housing stock. Furthermore, we have included a chapter on Australia, since social housing management there is affected by similar developments as those in Western-European countries and asset management has been relatively well-developed by the Australian social housing providers (Larkin, 2000).

To ensure a reliable result, all country monographs have been written by national housing researchers or experts of the local housing system. To ensure a comparable result, all contributors have written their chapter on the basis of the same format and research questions (see Appendix A). Basically, this format consists of four sections. The first section describes the social landlords that are operating in the concerning country. The second section discusses the national context for asset management in the social rented sector and deals with issues such as regulations, supervision, financial support and the housing market. The third section analyses the landlords' asset management practice: how do they determine their rent, allocation, sale, maintenance and renewal policies? Finally, conclusions are drawn and a critical appraisal is made of current practice in each country. The country monographs are presented in alphabetical order.

REFERENCES

Aaker, D.A. (1998) Developing business strategies (fifth edition), New York, John Wiley & Sons.

Ansoff, H. Igor (1984) Implementing strategic management, New Jersey, Prentice Hall International.

Boelhouwer, P (ed.) (1997) Financing the social rented sector in Western Europe, Housing and Urban Policy Studies, 13, Delft, Delft University Press.

Boelhouwer, P. (1999) International comparison of social housing management in Western Europe, Netherlands Journal of Housing and the Built Environment, 14, 225–240.

Boelhouwer, P., Heijden, H. van der and Ven, B. van de (1997) Management of social rented housing in Western Europe, Housing Studies, 12, 509–529.

Broeke, R.A. van den (1998) Strategisch voorraadbeleid van woningcorporaties: informatievoorziening en instrumenten, Delft, Delft University Press.

Bryson, J. (1995) Strategic planning for public and non-profit organisations, San Francisco: Josey-Bass.

Esping-Andersen G. (1990) The three worlds of welfare capitalism, Cambridge, Policy Press.

Fraser D.L. and Stupak, R.J. (2002) A synthesis of the strategic planning process with the principles of andragogy: learning, leading and linking, International Journal of Public Administration, 25, pp. 1199–1220.

Gruis, V. and Nieboer, N. (2001), What's strategic about housing management? A framework for comparative research, Paper presented at the ENHR-conference "Housing and Urban Development in New Europe", Puttusk.

Gruis V. and Thomas A. (2002) Approaches to option appraisal in social housing investement; English and Dutch experiences compared, Paper presented at the ENHR conference "Housing cultures; convergence and diversity", Vienna.

Gruis, V., Nieboer, N. and Thomas, A. (2004) Strategic asset management in the social rented sector; approaches of Dutch and English housing associations, Urban Studies, 41, 1229–1248.
Haffner, M.E.A. (1998) Housing statistics in the European Union 1998, Brussels, European Commission.
Hannagan, T.J. (1992) Marketing for the non-profit sector, London, Macmillan Press.
Harloe, M. (1995) The people's home? Social rented housing in Europe and America, Oxford, Blackwell Publishers.
Heijden, H. van der (2002) Social rented housing in Western Europe: developments and expectations, Urban Studies, 39, 327–340.
Hoekstra, J.S.C.M. and Reitsma, A.A. (2002) De zorg voor het wonen; volkshuisvesting en verzorgingsstaat in Nederland en België, Delft, Delft University Press.
Kemeny, J. (1995) From public housing to the social market; rental policy strategies in comparative perspective, London, Routledge.
Kemeny, J. (2001) Comparative housing and welfare: theorising the relationship, Journal of Housing and the Built Environment, 16, 53–70.
Kleinman, M. (1996) Housing, welfare and the state in Europe: a comparative analysis of Britain, France and Germany, Cheltenham (England) / Brookfield (Vermont), Edward Elgar.
Kotler, Ph. (1997) Marketing management (ninth edition), New Jersey, Prentice Hall International.
Larkin, A. (2000) Asset management strategies; a review of asset management strategies of housing associations in England and social housing providers in Australia, London, Metropolitan Housing Trust / The Housing Corporation.
Miles, R.E. and Snow, C.C. (1978) Organisational strategy, structure and process, New-York, McGraw Hill.
Mintzberg, H. (1994) The rise and fall of strategic planning, New York, The Free Press.
Müller-Armack, A. (1989) The meaning of social market economy, in: Peacock, A. and Willgerodt H. (eds.) Germany's Social Market Economy, London, Macmillan.
Nieboer N. and Gruis, V. (2002) What's strategic in the asset management of Dutch landlords, Paper presented at the ENHR conference "Housing cultures; convergence and diversity", Vienna.
Pearl, M. (1997) Social housing management; a critical appraisal of housing practice, London, Macmillan Press.
Priemus, H., Dieleman, F. and Clapham, D. (1999) Current developments in social housing management, Netherlands Journal of Housing and the Built Environment, 14, 211–224.
Scott, S. , Currie, H., Fitzpatrick, S., Pawson, H., Kintrea, K., Rosengard, A., and Tate, J. (2000) Good practice in housing management: a review of the literature, Edinburgh, The Scottish Exetutive Central Research Unit.
Smith, J. and Oxley, M. (1997) Housing investment and social housing: European Comparisons, Housing Studies, 12, 489–507.
Thomsen, A. and van der Flier, K. (2002), Updating the housing stock; the need for renovation-based approaches, Paper presented at the ENHR-conference "Housing cultures; convergence and diversity", Vienna.
Walker, R.M. and Zon, F. van der (2000) Measuring the performance of social housing organisations in England and the Netherlands: a policy review and research agenda, Journal of Housing and the Built Environment, 15, 183–194.
Willgerodt, H. and Peacock, A. (1989) German liberalism and economic revival, in: Peacock, A. and Willgerodt H. (eds.) Germany's social market economy, London, Macmillan.

2 AUSTRALIA

JOE FLOOD

2.1 SOCIAL LANDLORDS IN AUSTRALIA

Australia began in 1901 as a federation of six former penal and agricultural British colonies. Today following a number of waves of post-war immigration it has about 20 million inhabitants of diverse ethnic backgrounds. It is one of the most urbanised countries in the world: the five larger State capitals are primate cities with at least 60 percent of the State population in their boundaries. Much of the huge interior is unsuited for settlement and is very sparsely populated. Special housing responses have been necessary for people living in these remote areas, especially the indigenous inhabitants, the Aboriginal and Torres Strait Islander (ATSI) people. The housing and re-housing of servicemen following two world wars has also played a significant role in housing policy.

Australia is primarily a nation of home owners (see Table 2.1), with about 70 percent of households owning or buying their houses. The percentage of public stock is small, ranging from 4 percent in Victoria to 12 percent in South Australia. Most low-income people live in the private rental sector, which has about 23 percent of the total stock. The total public stock is around 400,000 dwellings, valued at about AU$30 billion (E17.5 billion).

Under the Constitution the States are equal partners with the Commonwealth, and retain primary responsibility for housing. Commonwealth involvement in social housing began in 1945 with the first Commonwealth-State Housing Agreement (CSHA), which has generally been renegotiated at five year intervals. Under the Agreement, the Commonwealth provides around $800 million for housing annually to the States,[1] which they partially match from their own sources.

Each State (and two Territories) manages its own stock under broad general guidelines negotiated in the CSHA, in organisations that fall under one or more government ministry but are known collectively as the State Housing Authorities (SHAs). These organisations are the eight major social housing providers in Australia. As well, they act as the strategic planning body in each

[1] From this must be subtracted about €220 million a year that the States repay to the Commonwealth for earlier loans.

Vincent Gruis and Nico Nieboer (eds.),
Asset Management in the Social Rented Sector, 19–37.

Table 2.1 Division of housing stock by tenure in Australia (2002)

	Social housing	Other rented	Owner-occupied
Percentage of total housing stock (approximately)	5.6%	23.8%	70.6%

Source: Australian Social Trends 2002, Australian Bureau of Statistics.

State or Territory for the provision of most forms of housing assistance, and as the funder and regulator of community housing. As well as public housing, they also have a large portfolio of low-cost housing loans and engage in other forms of housing assistance.

From the mid 1970s the Commonwealth began to provide fairly small amounts of "tied" funds to the States, for special purposes such as Crisis Housing, Community Housing and Aboriginal Housing. As Table 2.2 shows, about an eighth of the public stock is held in this way.

The various levels of government have also commonly provided housing in remote areas for employees such as teachers, policemen and nurses. This stock is managed under a variety of different arrangements in different States. Employee housing has declined in importance but is still significant, as Table 2.2 shows. The Defence Housing Authority in particular is the second largest landlord in the country (after the State of New South Wales).

The community housing movement has taken root under a variety of arrangements, following the international lead. The 24,300 dwellings designated as Community Housing are managed and/or owned by as many as a thousand different Community Housing Organisations, constituted as:

Table 2.2 Number of dwellings and managing agencies by sector, June 2000

Sector	Dwellings	Agencies[1]
Public housing	362,967	8
Community housing	24,316	1,000
Crisis housing	5,507	1,183
ATSI housing[2]	23,490	750
Government employee housing[3]	25,000	25
Defence housing	17,000	1

Source: Assembled from DFCS (2002) and various online resources.
Notes: [1] Approximate only. Includes managing agencies, not just owners.
[2] About 11,300 are in 700 Indigenous Community Housing Organisations (ICHOs) and the remainder is under the CSHA in various arrangements
[3] Approximate only. Firm numbers could be obtained for only four States.

- Community Housing Associations on the European model, with professional management. Some of these are very innovative and activist. In New South Wales, there are 44 CHAs ranging in size from 30 to 800 houses;
- Cooperative housing, with low-income tenant management. These are usually much smaller, with 5 to 30 houses. In Victoria they are run under a Common Equity Rental Cooperative umbrella, where a single organisation owns the houses managed by 100 co-operatives;
- Church housing. Churches own a good deal of urban land, and have been sympathetic to converting this land for low-cost housing. In NSW a single organisation now coordinates the setting up of church CHOs and constructs the stock.

Indigenous housing within and outside the CSHA is also mostly run according to one of these community housing models.

2.2 NATIONAL CONTEXT FOR ASSET MANAGEMENT

The key challenge for the asset management of Australian social landlords is to deal with an ageing housing stock in the context of a residualised social housing sector (with mainly poor clientele) and insufficient financial resources.

From 1977 pressure was placed on housing authorities to target public housing only to poor people, and to charge market rents to those who could afford them. Increasing social inequality from that time caused demand for public housing to increase much more rapidly than acquisition could take place, and entry increasingly was restricted to the most needy households. Today about 90 percent of public housing tenants are welfare recipients with low incomes. Public housing entry is now heavily prioritised towards people who are not only low income earners, but who are also in housing crisis situations, are disabled or have other special needs.[2]

A further impetus toward asset management is that much of the stock is now at the end of its economic life or it does not meet current needs. Concentrations of entrenched social disadvantage have developed in some public housing estates. Replacing this stock on-site or through disposal and acquisition elsewhere have become legitimate concerns of SHAs, although they were never considered in the past because of the dislocation to tenants.

---

[2] Only in South Australia where both sides of parliament supported public housing in the post-war years, and in the Territories where the dominant employer was the Commonwealth Government, were sufficiently large stocks built up to attract middle income earners. Today, SA and the ACT have about 18 percent of tenants paying market rents.

2.2.1 Regulations

The bilateral agreements of the 1999 CSHA require the States to:

- assist those whose needs for appropriate housing cannot be met by the private market;
- provide assistance for the duration of need;
- promote alternative methods of housing provision;
- ensure that the costs of assistance are transparent;
- provide assistance in a way that minimises workforce disincentives (DFCS, 2002).

Under the CSHA the States are required to charge market rents to higher income tenants. These are regularly reviewed. However about 90 percent of tenants are 'rebated', which means they pay no more than 20 percent of their income as rent. This means that the SHAs cannot vary rents received to any extent. The CSHA also requires the SHAs to obtain a replacement dwelling for any that have been sold or otherwise disposed of,[3] which must be at market value. Otherwise, the SHAs manage the stock as they see fit, subject to State policy.

2.2.2 Supervision

The SHAs are supervised on the basis of the CSHA. However, the CSHA was held in abeyance for much of the 1990s since agreement could not be reached on a proposal to cancel Federal capital funding and provide recurrent housing allowances to government tenants instead. The interim 1999 CSHA and the new 2003 CSHA have retained the existing system, offering somewhat reduced funding to the States, with part of it withheld unless they can demonstrate they have acted to reduce "poverty traps" or workforce disincentives, and to increase private sector involvement in social housing provision.[4] Other proposed 'reforms' are to increase the proportion of income paid by tenants as rent, and to negotiate term leases so that middle-income tenants can be removed from the stock.

Community housing organisations are accredited and monitored by the States, but have been otherwise largely unregulated. The Community Housing Federation of Australia[5] acts as a peak body for the sector and there are equivalent community housing federations in each State. These bodies are responsible for networking, encouraging good practice and service standards.

---

[3] This requirement seems to have been abandoned in recent years.

[4] Commonwealth Budget 2003, http://www.chp.org.au/Miscellaneous/FamilyandCommunity-ServicesCommonwealthBudget2003.doc

[5] www.chfa.com.au

Some States are developing a regulatory framework for community housing.[6] National Community Housing Standards are currently being negotiated and adopted. According to the 1999 CSHA, the challenge for the community housing sector is to prove that it is both viable and responsive to need.

2.2.3 Financial Support

Funds come from two main government sources: the Commonwealth which provides capital funding to the States, formerly as loans but now as grants, and circulating funds from the existing asset base and from repayments from the portfolio of earlier concessional loans to home owners. In the late 1980s, as lower income people increasingly came to dominate the stock, substantial recurrent deficits began to be recorded and the capital funding of the CSHA was increasingly used to meet the deficit, while money began to be borrowed from private sources in some States. This threatened the sustainability of the system and had to be discontinued.

In the neoliberal nineties, the amount of funds provided for public housing has fallen by about 30 percent in real terms, in favour of rent allowances provided to private renters.[7] This has forced the States to "do more with less". Acquisitions have become very few and the focus has moved to repairing the existing stock and managing it better.

A major problem with financial sustainability of the SHAs is that they are not funded in a way that matches their cash outflows. Rather than having the rental deficit explicitly funded, they receive grants which may vary from year to year, and revolving funds from a whole range of other activities, which means that it is difficult to determine the actual financial position of tenancy management and under what conditions the SHAs will have to sell off their stock to meet this deficit. It seems that several SHAs, particularly those with more stock like South Australia, have already reached this stage.

2.2.4 Housing Market

Australia's public housing sector is generally typified by a substantial housing shortage, resulting in long waiting periods. Because the social housing stock is so small, a considerable majority of poorer people have to find rental accommodation in the private market, where they are paying a high proportion

[6] See, for example, http://www.housing.qld.gov.au/strategic_policy/legislation/expl_chr.htm, or Clough *et al.* (2003).

[7] The Commonwealth Rent Assistance Program mostly pays a fixed weekly allowance depending on family type to about 940,000 low income people in private rental tenure, through the Social Security system. Outlays have more than quadrupled since it was first introduced in its current form in 1985 and are now about AU$1.9 billion (E1.2 billion) annually.

of their incomes as rent, sometimes more than 50 per cent; while they generally do not enjoy secure tenure as in public housing. It is generally agreed in housing policy circles (but not in general government circles) that all those on rent allowances (about 940,000 people) would be better accommodated in public housing. It would therefore require a doubling of the stock to house even these poorest members of the community adequately.

Some of the main issues that have been facing the SHAs are:[8]

- Changing demographics, including the growth of smaller households, family break-up, the decline of rural areas, and larger numbers of older households, which all require significant changes to the stock;[9]
- Greater social inequality, mostly resulting from 'casualisation' of the labour market and other neoliberal trade and capital account policies, that have put steadily increasing pressure on all forms of social welfare;
- Changing client base. Because most new tenants have special needs or are in a housing crisis situation, liaison must increasingly be made with social support agencies and extra costs of tenancy management are evident.
- Ageing housing stock. Over the past 50 years, the focus in public housing has been on expansion, generally at the expense of maintenance. The result is that States are now having to spend an increasingly significant share of capital funds to maintain and upgrade existing stock, parts of which were poorly and inappropriately designed and built.
- Dispersal of stock. All States have given some priority to addressing the problems of large public housing estates—particularly where design and location problems have contributed to significant social problems. Addressing these issues has often involved significant redevelopment and/or the break-up of the public housing estates, using mixed tenure strategies.

2.3 ASSET MANAGEMENT PRACTICE

Each State (and to some extent, each landlord) has its own approach to asset management, which depends on their stock, local conditions, historical factors, and the policies introduced by successive State governments. All have now taken on board the necessity for asset management, rather than past *ad hoc* approaches that involved a scramble for acquisitions, on-call repair, allocation by convenience and client management by fiat, though the extent to

[8] Drawn from Larkin (2000) and recent CSHA reports

[9] In South Australia 86 percent of new applications are now from singles or single parents (SAHT, 2003).

which formal asset management is actually being practiced differs markedly between sectors and jurisdictions.

It is also true that many practitioners do not understand exactly what asset management is, or take a quite narrow view. In the early 1990s most asset management strategies were concerned largely with physical renewal. Recognition of the need to integrate social and economic concerns grew in the 1990s, and asset management is now part of "whole of government" or "place management" strategies.

The SHAs have now become considerably more professional in their asset management practices. Prior to the 1990s, poor records were kept of the properties, tenants and waiting lists through primitive database systems that in some cases could not be altered once a record had been entered, and asset management was conducted through short-term bureaucratic procedures rather than any longer term strategy. This matched the expediencies associated with short government election cycle (typically three years) that made it extremely difficult for any class of infrastructure to be managed on the longer time frames required. Housing was never regarded as an asset that might be managed or sold, but as a one-off budget cost. For that reason, State Treasuries were exceedingly reluctant to spend on housing, regarding it as a Commonwealth welfare activity.

Until recently, the housing authorities have only been marginally interested in broader housing market issues. This is because their stock is too small to have much effect on broader markets, and because the market conditions they face are always the same—extreme shortage, with waiting lists of up to five years. With some SHAs now choosing to rationalise or even sell a portion of their stock, and with joint ventures becoming a common practice, market conditions are likely to occupy more prominence in management deliberations. From the mid 1980s the winds of managerialism swept through public housing, at the same time as they were being transformed into social welfare organisations. The new class of intermediate manager was often female, generally trained in social welfare or public administration rather than civil engineering, and client-oriented rather than stock-oriented. The result has been that direct bricks-and-mortar concerns have moved to the background in favour of financial management and client management approaches. These are exactly the required skills in a situation of steady or shrinking stocks and a client base with special needs requiring case management or other interventions.

2.3.1 Rent Setting

Following the Henderson Report on Poverty in 1977, States were encouraged to charge market rents and target poor households, making subsidies explicit. For most of the 1980s, the SHAs in fact attempted to implement a cost-rent formula for public housing which was never fully resolved, and all States eventually reverted to a "market rent" formulation. Nominal rents steadily rose until a reasonable approximation to market rents was "charged".

In practice these formulae were largely irrelevant both to the housing authorities and the tenants, since almost all the tenants were "rebated" or subsidised and paid no more than 20 percent of their incomes to the landlord. By the late 1980s this was no longer enough to meet even the immediate costs of repairs, administration, council rates and interest payments on past Commonwealth loans. The States were therefore forced to use an increasing proportion of their capital allocation to meet the recurrent shortfall.

Community housing also has very low-income tenants, with the vast majority on less than 50 percent of average household income, and 75 percent on statutory incomes. However the sector has been rather more able to manoeuvre since rents are more typically 25 percent of tenant income, possibly providing a small excess for capital purposes. A few housing associations and co-operatives have therefore made limited use of mortgage finance in order to purchase equity or increase their housing stock.[10] There has been a considerable debate as to whether tenants of community housing run under the CSHA are eligible for (private rental) Rent Assistance. This is extremely valuable since the landlords can make use of this to meet the operating deficit while meeting affordability standards for the tenants.

2.3.2 Allocations

Eligibility for the waiting lists is based on household income limits, which vary by State.[11] Because most waiting lists are 3 to 5 years long, in practice allocation is made on a two-tier priority system that in most States gives preference on a case by case basis to people who are in an emergency situation, have disabilities or are homeless.

The actual allocation of dwellings is commonly achieved by giving a client three choices related to their family circumstances and preferred location zone,[12] and if none are acceptable the client goes back to the bottom of the list. Transfers of existing tenants within the stock are also quite common.

Client eligibility for community housing is generally the same as for public housing, but different interest groups may add their own selection procedures for membership (e.g. women, disabled, aged, students, particular local government areas etc).

---

[10] Brian Elton and Associates (2000), http://www.nchf.org.au/finance.htm

[11] DHCS (2002), Appendix G. South Australia with its larger stock has much higher income limits.

[12] In New South Wales, clients must accept the first allocation.

2.3.3 Sales

From 1956–1972 under conservative governments, significant proportions of the better-quality and better-located stock were sold off to sitting tenants at a discount (Hayward, 1996). Since that time, the CSHA rule has been that property must be sold at market value. Sales to tenants are supported, at market value less the value of any improvements made by the tenants, and are exempt from stamp duty.

Current circumstances have encouraged some housing authorities to adopt private-sector methods in rationalising stock, selling off less useful stock or stock in excess supply (such as three bedroom detached housing, or housing in declining rural areas or where job opportunities have failed to materialise) and replacing it with housing more suited to client demand. Sales are also used as a means of diversifying the social situation in large estates. Overall, sales are running at about 6,800 per year or about 2 percent of the stock, slightly more than acquisitions. In a number of cases stock has been divested to the community sector.

Some public housing design styles have turned out to be inappropriate to the new reality of welfare housing. For example, Radburn-style estates popular in the 1970s with dwellings facing cul-de-sacs, and footpaths connecting houses on public open space, were intended to engender a pleasant community with reduced traffic risk. As the need to house priority cases has increased, these areas have experienced an increasing concentration of disadvantaged households. This planning design has turned out be inappropriate in a situation of uniformly low income and social breakdown, since poorly monitored public areas may become dangerous, and it is difficult to introduce home owners into the area to obtain a better social mix. In order to sell part of the stock to private buyers, it has been necessary to reconfigure road and dwelling layouts to conventional block form, to obtain separate titles for the dwellings, and to sell or otherwise privatise redundant public open space.

South Australia: Sale as Crucial Feature in Asset Management[13]

In South Australia, which developed using a manufacturing industry base, public housing started as "worker housing" in 1937 and was integrated with various government enterprises, including factories and shopping centres. A substantial stock was acquired in the post-war years, so that something closer to a European-style social housing sector was developed, occupied by a partly middle class clientele. Because the sector was seen as a genuine

[13] SAHT (2003), SGS (1997) Appendix 2, Larkin (2000).

alternative tenure, eligibility requirements for entry were much less stringent than elsewhere and waiting lists were much longer.

From the mid-1970s the public enterprises were divested. The loss of manufacturing employment meant that the State underwent a long-drawn out "rust belt" recession and poverty trebled. The SA government borrowed substantial funds at market rates for public housing to try to improve and increase the stock. Because Commonwealth funding in the 1990s came to be provided proportionally to population rather than the size of the housing stock, SA was hit particularly hard. Federal funding reduced by 60 percent in real terms over the past ten years, and rent received from increasingly poor tenants also fell by 25 percent in real terms, with both trends expected to continue. By the mid 1990s the rapidly increasing deficit per dwelling meant that action had to be taken to remove stock that was unpopular, of poor quality or contributing to social problems, and to apply the proceeds to obtaining better housing, to reduce the deficit and retire debt.[14] A plan was put forward to reduce the size of the stock from 63,000 to 42,000 dwellings.

The situation was treated as a challenge rather than a crisis. A key element in this strategy was the use of computer based options appraisal tools to determine what strategy to apply to particular groups of stock—to retain, renovate, sell as is, upgrade and sell, demolish and subdivide, or joint venture. At the project level quite detailed financial analyses are done, with outputs including net present value, internal rate of return and return on equity. Normally projects are self-funding through sales. A policy published in 1996 focuses on the rights of tenants affected, but it makes it quite clear that relocation will be required in some circumstances.

Some projects are:

1) At Mitchell Park, 9 km south of Adelaide, close to the Mitsubishi car factory, 75 percent of stock was in public ownership and the semi-detached double units were no longer in demand. Asset management problems of declining values, increasing maintenance costs and high vacancy rates needed to be addressed, and pressure on local social services was excessive. Following sale of most of the allotments, Community Perceptions Studies showed improved quality of life, while maintenance costs were reduced from €484 per dwelling per annum down to €135.
2) North Salisbury, a suburb 20 km north of central Adelaide developed to service engineering firms and a large arms factory, has 40 percent

[14] The Common Public Borrowing high cost debt was finally retired in 1999 (SAHT, 2001). The stock has already been reduced to 49,000 (SAHT, 2003).

public housing and the highest deprivation level in the State. The South Australian Housing Trust is implementing a master plan to reduce the concentration of public housing to 15 percent by demolishing poor quality semi-detached houses and blocks of units and selling the sites to five private developers in stages. The local government is cooperating by upgrading roads and community facilities.

3) The Parks area, 7 km from Adelaide, is one of the largest urban renewal projects in Australia. A formal joint venture has been established with Urban Pacific, a major developer. All but 450 of 2,840 public dwellings, mostly dating from the immediate post-war era, will be demolished, and housing density will be increased. To make the project self-funding, 3.5 old dwellings will be swapped for one new one. Risk is shared, so that the Trust will receive a proportion of increasing land values. Social and employment/skills development programs will be provided, and the existing strong community networks and the impact of relocations will be taken into account.

2.3.4 Maintenance and Renewal

Maintenance is still mostly conducted in the traditional manner, in response to direct demand from individual tenants or under longer term regular replacement schedules. The latter has been made easier by the construction of full asset registers that make it possible to prepare for replacement of various components well in advance, and on an estate basis. Optimal maintenance/replacement analysis is not widely conducted, except in the context of estate renewal—although there is now greater awareness of the possibilities (see examples of practice in section 3.3.5).

In some estates, tenant committees have been formed that manage the local maintenance budget in accordance with their ownpriorities.

The real problem is that in the past, a long-term life cycle approach to the stock was not applied and funds accumulated for future repairs are quite inadequate. This means that in some States, the whole capital budget is not sufficient to meet projected repair costs. Estate renewal has become the major focus of activity (as is illustrated in the examples from practice in the boxes).

Increasingly, the life cycle performance of the underlying asset base has come under scrutiny, with more formal means of assessing stock condition and the timing of various maintenance strategies or other options coming into use.[15] Most landlords are seeking to realign maintenance expenditure from

[15] See for example Tucker *et al.* (1996) or http://www.housing.qld.gov.au/builders/psi/psi_data.htm for the development of a formal Stock Condition Index based on a detailed analysis of various building components.

responsive to preventive and programmed maintenance, and also to reduce the maintenance bill by replacing old housing with newer, better quality housing more in line with modern community standards.

Community housing is generally newer and does not face the maintenance challenges of the older-generation public stock. Tenants also often pay bonds, which can be helpful in preserving property condition. However, ATSI housing suffers extremely high levels of damage due to high occupancy rates and cultural differences in the use of housing (with the stock typically depreciating at around 10 percent per annum) and a great deal of effort has been put into developing design, repair and maintenance strategies for the sector (see Flood, 2002 and SGS/Urban Resources, 1999). The community-based housing sector for older people is also facing a major asset management crisis, since like public housing, retirement villages were built using government capital funding with no provision for depreciation and major repairs.

Defence housing has been criticised in the past as being of poor condition, and the Defence Service Homes Authority has also been engaged in an extensive programme of upgrading and rationalising its stock using a similar approach to the SHAs.

2.3.5 Decision-Making Process: Considerations, Methods and Strategies

The SHAs are large, autonomous organisations and have had access to significant resources and expertise for planning and management, as well as access to research funding. It is in the areas of formal strategic management and asset management that most of the advances of the last decade have been made. Asset management has been recognised as the key to successful handling of the stock, and all States have set up sophisticated formal asset management mechanisms, which may include business planning, options appraisal systems, stock condition assessment, asset registers, online property management systems, planned maintenance schedules, and risk assessment.

Asset Management and Option Appraisal in New South Wales[16]

New South Wales (NSW), easily the largest landlord with about 127,500 dwellings, has always tended to be the most active of the States in finding alternative solutions to the financial and asset problem. As the "global city" Sydney is the most prone to residential segregation and has the highest house prices. The SHA has been criticised in the past for establishing large estates on the edge of town which now have some of the worst social

[16] SGS (1997).

problems.[17] The stock is considered inappropriate in location, quality and form (in particular too many three bedroom family houses built for conditions of the 1950s to 1970s). NSW therefore has good reason to consider how best to use its stock in the different parts of the city, and how to balance acquisition with maintenance and renewal strategies.

Overall, the major strategies that have been adopted in NSW are:

a) All new capital funds are to be spent on upgrading the existing stock to the point where it is a genuine marketable asset and provides good quality accommodation.
b) There are tentative moves to place the more problematic areas that require higher levels of social intervention into community—based housing associations.
c) There has always been a shortage of inner city housing and housing in more expensive locations. Existing sites will be developed to higher densities.

NSW has developed an 'Options Appraisal System' that is generally considered to be the most comprehensive formalised evaluation system in place. The options appraisal analysis is undertaken in two parts which are brought together to provide the overall assessment of the relative worth of different approaches or options to renewal: First, identification of issues, goals and available alternatives is achieved through widespread consultation and the agreement of all the relevant arms of the organisation. Then, a cost-benefit assessment is made of alternative strategies, making use of data such as:

- average value of dwelling;
- average value after improvements;
- market rents;
- acquisition costs;
- improvement costs;
- specific costs relating to such things as road works, titling, disposal etc.

Sensitivity analysis is used to ensure that the impact of alternative assumptions is fully understood. The output of this stage of the analysis is a set of conventional economic evaluation measures. Finally, the assessment of

[17] The Department of Housing actually moved its operational offices to Liverpool near the western urban fringe about 15 years ago, as part of a general State decentralisation policy but also because the majority of the clientele are in the west.

policy goals is designed to measure the extent to which outcomes will meet the Department's policy goals. The bringing together of the two strands ensures that the overall assessment measures goal achievement in the light of the efficiency with which it is achieved.

In addition to this system for option appraisal, NSW is currently developing a commercial investment system, a strategic dwelling performance measurement system , local area plans, condition assessment, and programme and project appraisal systems.

The New South Wales approach to stock management is interesting in that to some extent it seeks to emulate the operation of the private market within the public system. Under market forces, more desirable areas have higher land prices and therefore are built to higher densities. By redeveloping high-cost areas at higher densities, the SHA is duplicating the effects of the market without the intervention of a pricing mechanism. It also means that more intensive use is made of the high valued sites, with more rents being received. So far, other States have not been so market-oriented in their approach to redevelopment, and are more concerned with reducing costs and client welfare.

All SHAs are making use of formal decision-making tools. For example, the smallest State housing authority, Housing Tasmania, with a stock of 13,200 dwellings, has an Asset Decision Making Model using a variety of data extracted from the Tasmanian Housing Information System to identify nonperforming properties. The Preliminary Asset Decision Making Model analysis identifies a broad range of properties that exhibit some or the full range of negative performance attributes, with these properties then subject of further assessment and analysis in order to prioritise disposal (FACS, 2003).

The situation is different in the other sectors. Some parts of the Community Housing Sector and the ATSI Housing sector have poor managerial capacity and have found themselves near bankruptcy. A strong emphasis has been placed on correcting the situation through training, through manuals describing the whole range of asset management activities, and by replicating good practice.[18] To improve performance, accreditation of Community Housing Organisations is practiced, National Awards for Excellence in Asset Management are conducted annually,[19] and a National Regulatory Framework is under negotiation.

Top-down management and data collection

A strong top-down emphasis on performance management has been evident in government in Australia in the last decade, particularly the use of performance

[18] See SGS and Urban Resources (1999).

[19] http://www.nchf.org.au/awards2001a.htm#ppha

indicators as a means of monitoring national goals. Data on the diverse CSHA programs had always been collected and aggregated as part of the reporting process to the Commonwealth, though numbers were seldom reliable. From the mid 1990s an attempt was made to develop and benchmark a set of performance indicators, based on a framework common to all government activities. Initially these indicators were rather clumsy, concentrating on a rather narrow set of internal efficiency procedures in delivering housing, and were not at all integrated into SHA decision making processes (Urban Resources and SGS, 1999; Burke and Hayward, 1999). They have been improved in recent years, and now deal among other aspects with client satisfaction with the stock. To improve the situation, a National Housing Data Agreement has been signed with the Australian Institute of Health and Welfare to collect State data in an integrated way, and a National Housing Assistance Data Directory has been compiled to standardise definitions. (AIHW, 2002, 2003). Nevertheless the collection of these indicators is rather patchy; and standardised methods of calculating the indicators using the different State data collection and monitoring systems have yet to be firmly established. So far, the mechanism by which these indicators are to be articulated into actual policy change has yet to be determined. Also, bottom-up benchmarking between States to replicate best practice and to establish concordance of procedures is in its infancy.

Asset Management in Victoria

Victoria has had a rather more social welfare oriented public housing policy than New South Wales. Depending on the government in power, new initiatives in the 1980s focused on self help (such as urban homesteading) or empowerment (rental cooperatives). A large infill programme of building on inner sites, and a spot purchase program of buying existing housing rather than building new housing was tried from 1982, with the aim of achieving a better mix of stock. The latter had its own problems in that firstly the new stock generally required substantial and expensive refitting and rehabilitation, while upper limits on purchasing price meant that stock could only be bought in the cheaper suburbs. In fact spot purchased housing turned out to be just as spatially concentrated as the construction of housing estates.

Today the Office of Housing stock portfolio comprises over 72,000 rental dwellings with a total value in excess of AU$7.8 billion (€4.6 billion), supported by an annual capital program budget exceeding $400 million (€235 million). The Strategic Asset Management (SAM) Branch is responsible for setting capital program strategic directions and monitoring the performance of the social housing asset portfolio. Major functions include:

– Establishing a 10 year Asset Strategy to ensure the portfolio improves access to appropriate and affordable public and community managed housing;
– Within this 10 Year Asset Strategy, formulation of 4-year capital plans for acquisition, disposal, redevelopment, and upgrade programs;
– Assessing the viability and progress of all joint ventures and estate redevelopments and where those involve Office of Housing commercial interest, overseeing project management;
– Maintaining and developing electronic systems for management reporting and managing the capital program and property condition databases.

All rental properties are planned for inspection at least once every five years. A range of property condition elements are inspected and reported on. This information is recorded onto the Office of Housing Property Management System so that the Office of Housing can choose consistently from a variety of capital actions inclusive of acquisition, upgrade, redevelopment and disposal. A future trends assessment is conducted for stock being appraised, with stock categorised by local managers into 5 year bands in terms of likely future viability. Stock in the shortest time band will only be subject to responsive repairs and may be considered for sale or redevelopment.

The present property reporting system has been in place for seven years and it is now undergoing review with regard to improving data quality and timeliness, to establish better human resource capacity to meet inspection regime demands, and to develop strategic computer-based tools. Core computing systems are now under replacement to improve capacity. The new system under development is intended to have the following capabilities:[20]

– A property condition database with capacity to deploy data in decision-making regarding asset management strategies and programs;
– Property actions that will produce effective and affordable short and long term improvements in the condition of housing assets;
– Resource deployment decisions that will maximise the capacity to respond efficiently and effectively to trends revealed by property condition data;

[20] Tender document, *Asset Management System Victoria*. The tender was let to a British firm.

– Budget forecasting and budget allocations between often competing needs of property acquisition (future clients), and property maintenance (current clients) requirements;
– The formulation of short and long term upgrade and maintenance programs;
– The formulation of property disposal and asset re-investment programs, and
– The demonstration of compliance with legislative and policy requirements regarding risk management.

2.4 CONCLUSION

From the late 1980s the SHAs have been wedged in a situation of very high demand, diminishing Commonwealth and State financial support and reduced income from tenants. This has provided them with a substantial challenge in which more explicit and professional asset management practices and systems must play a major role.

Under neoliberal policies and international competitive pressures, the chances of increases in social housing funding in Australia are slim but demand will remain high or increase. Making use of the existing stock to its fullest extent and restricting occupancy to only the poorest and most marginalised tenants will be the only option for SHAs. Unfortunately the manner in which social housing is financed does not recognise the reality of the situation; in particular the necessity to fund ongoing and increasing rental deficits. Already parts of the stock are beginning to be sold to maintain financial viability rather than according to asset management principles of maintaining a quality stock that meets tenants' needs.

Most SHAs have adopted fairly comprehensive asset renewal programmes, strategies and systems in line with current practice, and often as a result of estate renewal plans from the early 1990s. Renewal strategies for highly disadvantaged local communities remain high on the agenda; the redevelopment of the physical environment is seen as a focal point for engaging the community in a process of development. Increasingly, social and economic criteria have been introduced into these plans as part of a "whole of government" or community development approach, as in other countries. However, it remains to be seen to what extent better management and planning of asset strategies can compensate for a lack of funding, or whether strategies can be sustained in the face of an always-volatile political agenda. The question also remains, as Randolph (2000) points out, how can we build communities and replace or displace them at the same time, which seems to be an inevitable result of asset-based strategies.

REFERENCES

Australian Institute of Health and Welfare (2002) National housing assistance data dictionary version 2, Canberra, AIHW.

Australian Institute of Health and Welfare (2003) Commonwealth-State housing agreement national data reports; public rental housing, Canberra, AIHW.

Berry, M, Flood, J., Lindfield, M. and Fisher, J. (1997) Institutional investment in housing: an Australian perspective; development of investment models aimed at encouraging institutional investment in rental housing; a report for the Department of Social Security, Melbourne AHURI.

Berry, M. and Hall, J., with the Allen Consulting Group (2001) Policy options for stimulating private investment in affordable housing across Australia; Stage 1 report: outlining the need for action; stage 2 report: identifying and evaluating the policy options (Prepared by AHURI for the Affordable Housing National Research Consortium).

Brian Elton & Associates (2000) Private finance and community housing: recent initiatives, paper for the National Community Housing Forum (May, 2000), http://www.nchf.org.au/finance.htm.

Burke, T and Hayward, D. (1999) Performance indicators and social housing in Australia, Institute for Social Research, Swinburne University.

Clough, R., Barbato, C., Farrar, A. and Phibbs, P. (2003). Stakeholder requirements for enabling regulatory arrangements for community housing in Australia., Australian Housing and Urban Research Institute, Sydney Research Centre http://www.ahuri.edu.au/attachments/pp_stakerequirements.pdf.

Department of Family and Community Services (2002) Housing Assistance Act annual report 1999–2000, Canberra, DFACS.

Flood, J. (2002) Modelling construction and repair decisions in indigenous housing. Paper for the European Network for Housing Research Conference, Vienna, July 2002.

Hayward, D. (1996) The reluctant landlords: the history of public housing in Australia, Urban Policy and Planning 14, 1, pp. 5–35.

Larkin, A. (2000) Asset Management Strategies; a review of Asset Management Strategies of Housing Associations in England and Social Housing Providers in Australia, London, Metropolitan Housing Trust/The Housing Corporation.

McNelis, S., Hayward, D. and Bisset, H. (2001) A private retail investment vehicle for the community housing sector: positioning paper, AHURI Swinburne-Monash Research Centre.

Randolph, W. (2000) A review of public housing estate renewal in Australia. A presentation to the Queensland Department of Housing Urban Renewal Seminar (April 2000), Urban Frontiers Program UWS.

South Australian Housing Trust (2001) Chronological History of the SAHT.

South Australian Housing Trust (2003), Strategic Asset Management Plan 2002–2007, Adelaide, SAHT http://www.housingtrust.sa.gov.au/text/text.asp?swId=1&snId=3.

SGS (1997) Public housing estate renewal in Australia; AHRF Project Number 212; (Appendix 2: case studies; Appendix 3: New South Wales housing options appraisal system; Appendix 4: manual for sectored cost benefit analysis).

SGS (1999) (with Urban Resources) Identification of strategic asset management best practice for indigenous housing organizations; Volume 1, Policy and Strategic Issues (June, 1999).
SGS (2000) Public housing estate renewal in Australia (AHRF Project 212).
Tucker, S.N. et al. (1996) Housing stock condition index (AHRF Project), CSIRO Division of Building, Construction and Engineering.
Urban Resources and SGS (1999) Methodology for developing benchmarking for housing outcomes (AHRF Project 208, Final Report).

3 AUSTRIA

EVA BAUER

3.1 SOCIAL LANDLORDS IN AUSTRIA

There are two main types of social landlords in Austria: municipalities and limited-profit housing associations ("Gemeinnützige Bauvereinigungen")[1]. Together they own about 870.000 dwellings which is about one-fifth of the total housing stock (see Figure 3.1). Ten years ago, the social rented sector the municipalities and housing associations each owned about 50% of the social rented housing stock, but the share of the municipalities has decreased in the last decade due to their withdrawal from new construction.

The social rented sector in Austria is not the only sector where special conditions create some distance to 'pure' market housing and thus may be headed under social housing. The Austrian Housing Promotion System also functions as financial aid to private investors and private persons for the building of rental housing, condominiums (owner-occupied flats, with shared ownership of land and common parts of the building) and single-family homes—of course with some restrictions regarding the rent level and the resale of the subsidised assets.

In total, about 50 percent of the total housing stock in Austria has been financed with some public help, including municipal housing that has been financed from the municipal budget.

In Austria, the municipalities are the 'older' generation of social landlords, having started their activities in a time when public subsidies to private entities were virtually non-existent. These public subsidies, which were an outcome of political agreement after the end of World War II, were an important base for limited-profit housing associations getting into housing production. From this time on, many of the municipalities switched from direct engagement in housing to engagement via housing associations. Only the City of Vienna took a different way, fully keeping on their own production of housing until the 1970s and slowly reducing it since then. That is why today in all federal

[1] The traditional translation into English would be 'non-profit' housing associations but, taking into account the regulations that we will discus below, 'limited-profit' is more accurate in our opinion.

Vincent Gruis and Nico Nieboer (eds.),
Asset Management in the Social Rented Sector, 39–59.

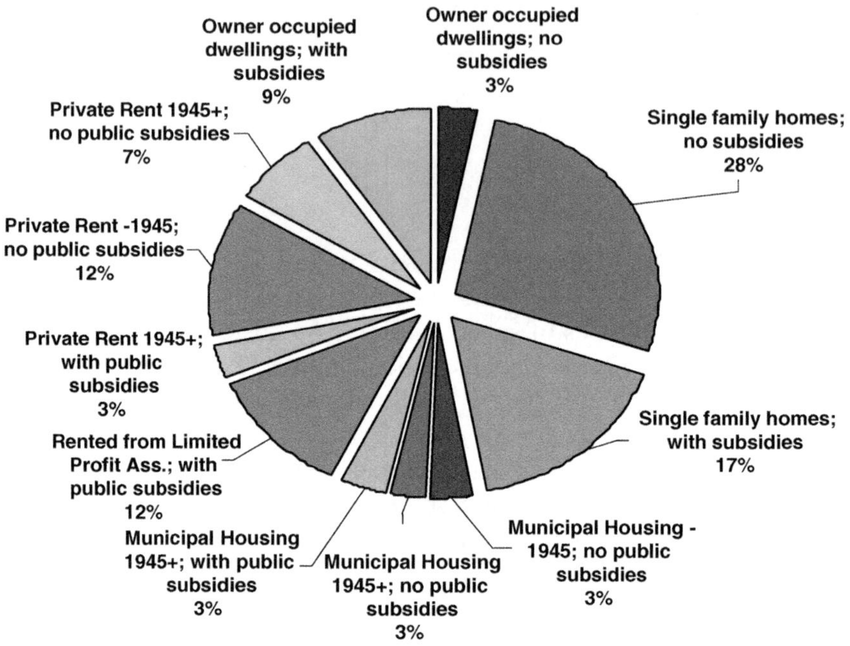

Figure 3.1 Estimation Austrian housing stock by tenure and public subsidisation (total 3.8 million dwellings) (2000)

provinces—except for Vienna—the social rented housing stock managed by housing associations exceeds that of the municipalities.

The history of social housing corresponds with the age structure of the social housing stock. While only 11% of the social rented stock owned by housing associations has been built before the year 1945, the respective share from the municipalities is one third; the dwellings built after 1980 only make up for 10 percent of the municipalities' housing stock but more than 40 percent of that of the housing associations' stock.

There are big differences in the size of the managed housing stock. There are more than 2,000 municipalities running social rented housing, but there are very few with a large stock. Except for Vienna there is no municipality which owns more than 10,000 dwellings. Furthermore, some municipalities have transferred the management of their housing stock (or parts of it) to housing associations, altogether about 10%. The average number of managed rental dwellings is very low in case of the municipalities—although the city of Vienna is one of the biggest landlords in Europe. The average number of the associations' rental stock is about 2,630, but this figure does not represent the entire field of activities as there are also dwellings in condominiums which are managed by the associations—the average sum of dwellings managed by an Austrian housing association is about 3,900.

Table 3.1 Some characteristics of social landlords in Austria (2000)

Characteristic	Municipalities	Housing associations[2]
Number of dwellings	355,670	494,800
Number of landlords	2,062	188
Dwellings per landlord (average)	172	2,632
% of total housing stock	9.6%	12.8%
% of social rented stock	43%	57%

Source: STATISTIK AUSTRIA/Österreichischer Verband gemeinnütziger Bauvereinigungen

Table 3.2 Social rented dwellings and associations by type of owner (shareholder)

Landlord	Owner	Number of dwellings	% of total number of dwellings	% of housing associations
Municipality	Municipality	370,000	42.8	—
Limited-profit housing associations of which the (majority) owner is...	Municipality	46,640	5.4	9.4
	Federal province	46,800	5.4	9.5
	State	60,000	6.9	12.1
	Association	34,090	3.9	6.9
	Employer	9,730	1.1	2.0
	Business company	19,470	2.3	3.9
	Bank/insurance company	52,500	6.1	10.6
	Cooperative	225,770	26.1	45.6
Total		865,000	100	100

Source: STATISTIK AUSTRIA/Österreichischer Verband gemeinnütziger Bauvereinigungen

To understand the character of Austrian housing associations it is important to discuss the different types of enterprises as well as the different types of owners (shareholders). About 50 percent of the associations are constituted as cooperatives, the others are limited companies and joint-stock corporations. Most of them look back on a long history—90% of the associations has been founded 40 and more years ago. Due to this long period, the changes in housing conditions and housing policies and the (consequently) transformed role and function of housing associations, the character of many associations has changed since the starting point of their activities. There have been changes in the structure of ownership as well as re-orientation in their 'business philosophy'. Table 3.2 shows the current structure of ownership of the still active housing associations.

[2] Including the five associations owned by the central state which were exempted from profit regulations in the year 2001.

3.2 NATIONAL CONTEXT FOR ASSET MANAGEMENT

3.2.1 Regulations

There is a comparatively complex set of regulations for social housing in Austria. The first regulations concerning non-profit housing are to be found in old acts concerning housing promotion and cooperatives in general (e.g. Lugger, 1994) It was a Nazi-German law that drew together the principles of non-profit housing in one legal act. In the year 1978 a new Austrian act for this sector was established in which the main principles of non-profit housing are defined (which does not mean that all regulations are to be found in that act; the exemption from corporation tax for example is laid down in the corporation tax act) (Holoubek, 1994). One of the very specific elements of Austrian housing policy is the promotion scheme through which the production of affordable housing is stimulated (see section 2.3 for details). Some of the most prominent developments within housing regulations during recent years were:

- liberalisation of (rent) regulations in private housing;
- a variety of amendments in the Limited-Profit Housing Act with no general trend—the introduction of new restrictions was combined with some deregulation;
- a shift in the responsibility (not in funding) for housing promotion from the state to the federal provinces.

General administrative regulations

The basic rules for housing in Austria are laid down in different laws, which together make up the framework for the activities of the social landlords (as well as for others):

- The Rent Act contains principles of rent contracts (tenants' and landlords' basic rights); rules for rent setting for private, non-subsidised housing (where "private" includes each dwelling outside the limited-profit sector);
- The Limited-Profit Housing Act ("Wohnungsgemeinnützigkeitsgesetz") contains specific rules for housing associations concerning for example compulsory re-investment of own funds into housing, rent setting according to principle of cost-coverage and the definition of a system of supervision (see below for details);
- The Home-Ownership Act which contains among others rules for the decision-making process within condominiums;
- The Housing Promotion Acts ("Wohnbauförderungsgesetz") which are made up by the federal provinces which are in charge of housing promotion (the central government is involved through the financial funding of housing promotion).

It is interesting to note that municipal housing is not the subject of a special law, while there is a specific act concerning limited-profit housing (associations). This Act includes:

- definitions of the 'accepted' activities of the housing associations which are recognized as limited-profit ones by the public authorities (federal provinces). The most important restriction is that housing associations have to restrict their investments to housing. Within housing, associations are further restricted because they may not build or administer flats exceeding a useful floor space of $150m^2$. For some activities—for example building activities for third parties—an agreement by the public authorities is required;
- principles concerning economy and functionality to prevent inefficient behaviour (such as installing a 'golden toilet'). To enable supervision of these principles the act defines income ceilings for the managers and limits for the costs of administration, planning and building;
- principles of rent and price setting (see the details below);
- limitations of profits for the shareholders and the level of interest on own funds;
- a system of public supervision on the basis of the principles mentioned above.

These principles leave room for activities besides social housing—such as construction of some more expensive (but still cost-price) rental or owner occupied dwellings, but in general the activities have to in the interest of society.

Rent setting

There are different rent regulations depending on:

- type of landlord (limited-profit association/other);
- age of the dwelling: there are different regulations for new (constructed after 1945) private and municipal housing and older buildings;
- subsidisation (Promotion Acts also contain principles for rent setting).

Table 3.3 gives a summary of the most important criteria for rent setting. It should be noted that there are no special legal regulations concerning the rents of municipal housing—the rent is set either according to the rental act or, in case of subsidisation, according to the principles stated in the Promotion Acts.

Regarding the further development of once negotiated rents also different principles can be identified:

Table 3.3 Overview of rent-setting criteria

Type of landlord, building and subsidisation	Regulations for rent setting			Non-legal influence on rent	
	Respective law	Principle of rent setting	Component maintenance and renewal?	Political	Economical
Municipality, non-subsidised, constructed before 1945	Rental Act (which applies to all types of landlords except housing associations)	Based on "Richtwert", which is a limited market rent	No separate rent component earmarked for maintenance and renewal	Political pressure/ decisions to remain below legal maximum	Market-conditions and socio-economic characteristics of applicants
Municipality, non-subsidised, constructed after 1945		Free (market) rent			
Municipality, subsidised	Promotion Acts	Cost-coverage (with small margins)	Earmarked rent component, defined by law	Amount of subsidisation	Market conditions concerning costs of land, construction and finance
Housing association	Limited-Profit Housing Act				

- Free market rents are subject to negotiation;
- For statutory rent limits the increase rates are defined by law (index-linked);
- Rents that are set according to the principle of cost coverage follow the dynamics of repayments for mortgages and/or annuity grants.

For the housing associations there are specific rental regulations based on the principle of cost-price renting with a certain margin for the associations. Based on the Limited-Profit Housing Act a maximum and a minimum allowable rent is defined. The minimum is determined by the costs; remaining under the cost level is not allowed. The definition of costs includes repayments and interest for (mortgage) loans as well as interest on own funds and is bound to the project. During the repayment period, housing associations are also allowed to earn a margin of 2 percent above the costs. When repayments are finished the rent remains at its former level or is reduced if it exceeds the maximum level which is defined by law (at a level below market rent). There is also a specific component in the rent that is earmarked for maintenance, repairs and renewal. A part of this component has to be paid back to the tenants if it has not been invested. This component is also exclusively bound to the project. Thus, the rent level is influenced by the historical costs of land, costs of construction, system of financing and the age of the building. Housing associations can influence (reduce) rents indirectly by reducing the costs, for example by replacing one mortgage loan by another or by investing own funds instead of loans.

In case of new leases there are specific rules. If the cost-rent level is below a certain level (the same level which defines the ceiling of the rent after the term of the mortgages) the new tenant may be charged this higher rent. This possibility to raise rents at turnover is particularly relevant for the housing stock that has been constructed before 1970, because the rent levels there are very low in this part of the stock.

Allocations

Regulations for allocation are laid down in different acts:

- The Housing Promotion Acts of the federal provinces contain income limits which limit the accessibility of subsidised housing. These limits vary between provinces—they are higher in the eastern provinces (including Vienna). These income ceilings are of major importance for housing policies and housing markets, since they define the room for building activities in both the limited-profit and the free sector;
- Municipalities create their own allocation schemes; either they have to accept the income limits set by the promotion acts or in case of

non-subsidised housing they set their own. In addition, they use criteria concerning the current housing situation of applicants as well as size of households. In general, the income ceilings of municipal housing are below those of subsidised housing;
- There are no specific rules concerning the allocations of housing associations, but they are obliged by the Limited-Profit Housing Act to use objective criteria for the letting or selling of dwellings. And, of course, they are obliged to apply the income limits according to the Housing Promotion Acts.

In addition to these formal regulations, there is 'informal' cooperation between municipalities and housing associations through capital involvement of municipalities and the provision of building land in return for the acceptance of tenants who are nominated by the municipality.

Sales

Owner occupancy in multi family blocks is one of the major issues in Austrian housing policy. Although owner-occupancy always has been a central goal of conservative and Christian housing policies it has been institutionalised in Austria only in the year 1948 since there has been much of scepticism against the co-ownership in multi-family blocks (Bauer, 1993a).

The regulations concerning owner-occupancy have been laid down in a specific act. Furthermore, the Limited-Profit Housing Act contains the right to buy to tenants who contribute a certain amount out of their own funds to co-finance the costs of land and/or construction. Such tenants' contributions have become very common in Austria. In general, tenants' contributions are treated similar to loans. When tenants move out of their dwelling, they get their money back, under reduction of a certain margin. Since the introduction of these regulations in the Limited-Profit-Act, the production of traditional housing for sale decreased from one-third to less than 20 percent in recent years. If the future sales to the tenants will compensate for this decrease in the production of owner-occupied dwellings is not known yet since, due to tax regulations, tenants have to wait for ten years before they can effectuate their right to buy. In the last two years there also seems also to be a shift from traditional renting to offering dwelling with the option to buy.

As mentioned above the limited-profit associations are also active in the production and sale of owner-occupied housing (condominiums). About one third of the managed housing stock are owner-occupied dwellings in condominiums. The Limited-Profit Housing Act also sets limits to the price of new dwellings via two different models. The traditional model is the cost-oriented price which is defined by cost of land, construction, planning and administration plus a 2-percent-margin for the associations. Some years ago, a 'fixed price' method has been introduced which allows an additional risk-margin for the associations in exchange fixing the price in advance (before actual costs are known).

In case of sales out of the existing rental stock there was no exact definition of the selling-price. In principle, the cost-oriented price principle had to be applied, but this was hard to determine, since the costs are related among others to the costs of land which value has changed since the building-year. Another question concerns the acceptable price for the tenants, who feel that there has to be a 'reasonable' price-reduction in order to take into account the rent that they have been paying for years. Furthermore, there is the issue of the subsidies that have been granted for the building. Some of the authorities ask for the (partial) repayment of granted subsidies in case of the sale of rental flats (as in general owner-occupation is less subsidised than rental housing). But, new regulations are in preparation to facilitate the calculation of reasonable sale-prices.

Maintenance and Renewal

The regulatory context for municipal housing and limited-profit housing concerning maintenance is different due to the different age of the housing stock, the different systems of rent setting and the concentration of low-income households.

Housing standards in Austria have been rather poor in the past. Even municipal housing, where great improvements were made compared to private housing, could not really be compared to standards in many other European cities. In the municipalities' rental buildings dating from the 1920s and 1930s and even in the first post-war buildings there were no bathrooms and no central heating systems. Instead there were common showers and laundries in the blocks. As the rents were kept very low there was no money for basic improvements. It was not until the establishment of the promotion system for housing renewal in the 1970s that the municipalities could start their basic improvements. Since then, one block after the other has been modernised and improved—the general parts of the buildings (including building in elevators) as well as the amenities in the dwellings. In case of improvements within the dwellings the agreement of the tenant is required (and some older persons do not agree). Nowadays, some of the municipal pre-war buildings are still waiting for these basic improvements. But, even the subsidisation cannot prevent the increase of rents in those buildings—although there are also individual allowances available for the low-income households which often are concentrated in those older buildings.

The situation in the limited-profit sector is much better as the housing stock is younger and there is a rent-component dedicated for renovation. Of course, also among housing associations there are some cases where their own funds are not sufficient to cover the costs of repair and/or improvement. In these cases the associations have to apply for public subsidies. Nevertheless, even in contrast to the latest energy standards, the housing stock of the associations shows a satisfactory condition in general. The introduction of the earmarked rent-component has contributed to this good relatively condition. Until the early 1980s the need for improving the housing stock had not been fulfilled in order to keep rents on a very low level. A housing reform introduced rent components—also for existing contracts—which were dedicated for repair and improvements. Today these components only exist in the limited-profit sector and subsidised housing. And, in case there are major repairs to be financed and the rents (including the earmarked components) are not sufficient to cover the costs there is the possibility of additional rent-increases for a limited period.

3.2.2 Supervision

Municipal Housing is subject to public control by the Audit Commission ("Rechnungshof")—a department that has been established at the central state's level with the function to control public entities, as well as in Vienna by the City of Vienna Control Department ("Kontrollamt der Stadt Wien") which functions in a similar way on the municipal level.

Housing associations are subject to a system of public supervision as well as a specific audit system which is based on the regulations for audition of cooperatives. Subject of audition are the general economic performance as well as the special regulations set by the Limited-Profit Act (e.g. Patera, 1986; Scherz, 1994). In practice, supervision plays an important role since, in practice, the rules defined in the Limited-Profit Housing Act (see above) give way for some interpretation. For example, there can be questions about efficiency if the management costs exceed the required limits while the general economic performance is (highly) positive and it is the system of supervision which has to raise these questions. Thus, the supervision plays an important role in defining the practical implications of the rules and the limits to the activities of the associations.

3.2.3 Financial Support

While in some other countries public support of social housing has been decreased or has shifted from direct, object subsidisation to individual allowances, in Austria object-subsidisation is still prominent (see Matznetter, 2002). About three thirds of new construction is partially financed by public loans or grants. There are also individual housing benefits—but only a small proportion of households receive these (between 3 and 4 percent) as there is the basic promotion scheme which keeps housing costs on a low level. Of course, there is a debate about reducing public funds for the purpose of housing promotion as well as to adapt the system in the direction of that of individual housing allowances, but a majority of the relevant parties still support the existing system very strongly.

The responsibility for housing subsidisation lies at the federal provinces while the majority of financial means comes out of the state's budget. The provinces have the responsibility to formulate the promotion schemes' targets, nominate the landlords/investors for the fulfilment of the schemes, define the way of subsidisation and indicate income ceilings for future tenants/owner-occupants of subsidized housing. As a result, there are nine different Promotion Acts with different principles which are subject to amendments every couple of years. The common principles and the most important differences are as follows:

- There are object subsidies for new construction of rental dwellings, condominiums and single-family homes as well as for maintenance and improvement of existing buildings and dwellings;
- Recipients of the subsidies for rental and owner-occupied housing are/have been municipalities, limited-profit associations and for-profit investors (since about ten years, also concerning subsidised rental dwellings in some of the federal provinces) and, in case of owner-occupied single-family homes, individual households as well;
- Subsidization of maintenance and renewal of buildings goes to all types of landlords and owners. However, for landlords to be eligible for subsidies for maintenance and renewal the lack of own means is a condition. This is why limited-profit housing associations seldom apply for public grants. There is one exception: in case of measures aimed at reducing the use of energy some provinces grant subsidies regardless of the financial situation to induce sustainable improvements.

Currently, there is very little new construction by housing associations without subsidisation in form of public loans for a—major—part of costs of construction (up to 70 percent of it) or grant to reduce the costs of financing. Consequently, the rent level is about half of the amount that it would be without public aid. This seems high, but it has to be taken into consideration that in other countries there is much more indirect support of housing, while this is of minor importance in Austria (see Matznetter, 2002). Now, discussion in Austria is focused on the further performance of the housing promotion scheme. During the year 2004 the respective contribution of the federal state has to be negotiated between the federal ministry of finance and the federal provinces and (of course) there are intentions to reduce this contribution.

There are differences in the financial position of the housing associations but generally they are in a financially healthy status. There are some complaints about the lack of own capital e.g. in comparison to associations in other countries. Those associations that have focused on the production of dwellings for sale have been in a better position regarding own funds and associations with a comparatively old rental housing stock tend to show a better financial performance as well. In the 1970s, 80s and 90s, some smaller associations have experienced financial problems and during that period a special tax law enabled the merger of smaller cooperatives. Today, lacking a special tax treatment, there are very few mergers. What does happen is the transfer of the management to bigger associations in order to optimise the cost/income ratio.

3.2.4 Housing Market

The conditions on the housing market have changed during the last decades. While up till the 1970s, due to historical reasons, there had been a long period of undersupply of affordable housing, especially in the bigger cities, the situation changed in the early 1980s after a period of substantive building activities. But, the then following decrease in new building activities led to a new housing shortage (e.g. Fassmann et al., 1992). This shortage was answered by building boom with a peak between 1996 and 1998. The average figure of 6,4 new constructed dwellings per 1.000 inhabitants between 1991 and 2000 was within the highest of the European Union (CIRIEC, 2003). Thus the situation on the Austrian housing market changed again and in some regions there seemed to be a slight oversupply. As the activities in new construction went down dramatically in the last three years—although they are still above the level of most other EU-countries—the housing market tightens again. As the results of the last housing survey held in 2001 are not yet published, there are no new estimations of the housing production for the coming years, but it seems that the actual level of production is below the required quantity. Furthermore, it is yet unclear what the consequences of migration after the EU-expansion will be. What is certain is that there is still a lack of affordable larger dwellings in the cities. Nevertheless the housing market in Austria has undergone a general change compared to the situation some 20 years ago since the quality of the housing stock has improved very much. The average size of the dwellings is still behind that of other countries but this has improved as well. That is the reason why the recent years created a situation in which there are more possibilities to choose and households in general live under much better conditions than some two decades ago. Consequently, households who decide to move do not take the first dwelling they are offered anymore.[3]

3.3 ASSET MANAGEMENT PRACTICE

Both municipal housing and limited-profit-housing undergo some basic changes regarding their (management)-philosophy and organisation of their institutions. Both type of landlords have to deal with an increase of wealth on one side and increasing differences in income and social opportunities on the other. Both are confronted with people seeking for accommodation in a situation of general greater choice. In reaction to these developments the municipality of Vienna—as mentioned one of the biggest landlords all over Europe—has restructured its organisation. The municipality has converted the administration of the dwellings from a public board to a private-like business enterprise ("Wiener Wohnen"), without giving up the social aims. The goals of this process are to reach more efficiency in the management process and better commercial results. Because of their tradition, the limited-profit associations already had more commercial-orientated structures, but due to the changed situation on the housing markets and the new competition—last but not least created by the acceptance of commercial developers of subsidised housing—there is a re-orientation to more business-like structures. Some of the associations have been sold by their former owners who have lost interest in housing due to a diminishing need (especially institutions producing dwellings for their employees) and the new owners' are focused more on better commercial results. Thus the asset management practice has gained more complexity in social housing, combining more business-like behaviour with social objectives.

3.3.1 Rent Setting

The rent setting practice of municipalities differs from that of the housing associations, among others due to their differences in their (political) background, their regulatory context and the composition of their housing stock and their tenants.

A prominent principle of rent setting within municipalities is that they aim to keep rents below the maximum that is allowed by law. This corresponds

[3] For example, young people who are looking for a dwelling of their own can now afford to be more picky since the dwelling they live in with their parents is now big enough to stay in for a little longer (while the dwelling their parents used to live in with their parents was small and lacked the basic amenities).

with the fact that municipality housing is dedicated to lower income households, but there is now discussion about changing this principle and at least to differentiate according to quality and location of the dwellings.

The principle of cost-coverage does not leave much room for housing associations to influence the rents after the model of financing is fixed. It is not possible for them to raise rents, even if the market situation would allow additional rent increases. Housing associations can lower their rents if they succeed in reducing the costs (e.g. by replacing loans by own funds). In practice, this often takes place since the methods of financing are largely influenced by the subsidisation method, which is often based on low rent levels at the beginning of the exploitation period with comparatively sharp increases afterwards. This does not only has a negative influence on affordability but also leads to a process of 'house-jumping' since rents in new buildings are lower, which has become a major problem in the past decade. These developments have led the housing associations to implement more sophisticated methods in monitoring the cost development in the rental buildings to be able to take the necessary steps at the right time, which have replaced the simple year by year calculation of rents. The strategies differ between associations. Some of them try to avoid every rent increase that is caused by an increase of the repayment of a loan by replacing them by own funds or another loan; others only do that in case there are conflicts with the rent level in other (newer) buildings in the same area. The competition from new dwellings also influences the rent setting at turnover when there is the possibility to raise the rent if the cost rent is below a certain level. Even if this level is very low it might occur in some regions that rents in new constructed subsidised buildings are still lower what could cause problems: either a 'real' problem of lettability or a harmful effect on the reputation of the landlord.

3.3.2 Allocations

Allocation of dwellings has become a major issue in the last years, since it is regarded as an important factor for social integration. Housing policies put more emphasis on this question and larger housing projects are now always planned as co-production between different landlords to get a social mixture of inhabitants. Although the problems around allocation are similar for all social landlords there is one important difference between allocation in municipal housing and limited-profit-housing, namely the impact of (general) politics.

As mentioned above the municipalities have nearly withdrawn from new construction. Thus, the question of allocation is concentrated at the actions concerning letting to new tenants in older buildings. One of the most prominent questions regarding allocations policies is the combination of social goals and the knowledge about the important role of integrating different social classes and ethnic groups. Especially in Vienna there is also much public interest in and discussion about the possible strategies. As a consequence, the strategies of Wiener Wohnen concerning allocation are still influenced by political decisions to a large extent.[4]

Of course there is also political influence on the decision-making process of housing associations, although the extent depends partly on the owner and/or the integration into the political network. Nevertheless, there is more room for a more rational practice. Another important difference with municipalities is that allocation of associations' dwellings concerns both the existing stock and new construction. The income limits of subsidised housing allow a broader target group, and thus a better social mix.

In the last years, due to the described developments in the market situation, the process of allocation has changed. Formerly, it was mainly a management of waiting lists. Today there are much more marketing activities and tenants are seen more and more as 'normal' customers. This is does not mean that the waiting-lists have disappeared, but people no longer accept the first dwelling offered, for example if they consider it to be too far away from the city center. Today the dwellings in more remote areas have to have some extra quality (like little private gardens, terraces or good architecture) and the environment has to be improved as well, which is not only the job of the housing associations but also that of other actors involved (investors/landlords, public authorities).

3.3.3 Sales

The issue of sale is often discussed, but the sale of rental units is not very popular among municipalities and associations. Within housing associations, housing for sale is dominated by the production of new housing (condominiums) and not of selling out the rental stock. There are not more than 12,000 rental dwellings which have been sold by the associations to their tenants within the last three decades, which is not more than 2–3 percent of the total rental stock. One reason for refusing to sell rental dwellings is that, among others due to the legislation, the management of buildings with mixed tenure is rather complicated. Furthermore, there is little interest among tenants to buy their dwellings, partly because of the rent system (cheap rents) and partly because of cultural influences. Sales of rental units mainly result from the financial situation and the need for liquid means, but this need is limited

[4] To give an example: Accepting immigrants in municipality housing is one of the most important issues in Viennese housing. Foreigners may not apply for municipal housing which is a political demand on one hand and a subject of criticism on the other. What happens is that the municipality assigns the Austrian Citizenship to immigrants in order to give them access to municipal housing.

(although, it is being considered to introduce incentives for selling the existing rental stock in order to get financial means for new construction).

Concerning new production, it depends on the political orientation and/or business strategy of the association whether housing production is concentrated on rent or owner occupation, although as a result of the introduction of the right to buy (and tenants' contributions) some 'specialists' gave up their focus on either production "classic" owner occupation or rent. The construction of dwellings with the option to buy has a financial advantage (since tenants contribute in the costs) and also has some influence on the social background of the tenants. But, again, the majority of social landlords fear the situation in which they have to manage a building with mixed ownership.

3.3.4 Maintenance and Renewal

As stated above, municipalities and housing associations have to act under very different conditions concerning maintenance and renewal. This is reflected in their strategies.

For the renewal and modernization of the older and weaker housing stock of the municipalities public subsidies are required. Due to financial restrictions the municipalities have to develop strategies for improving their housing stock step by step. The renewals also have an impact on allocation strategies as the rents often have to be raised to be able to finance the improvements. Consequently, the process of renewal requires much organisation and social engagement. As mentioned above the tenants are not forced to accept improvements within their dwellings but of course it makes sense to combine that with the modernization of the building. Thus tenants have to be persuaded to cooperate. Furthermore, during the renewal activities there are of course nuisances for the tenants and many complaints to deal with. The municipality of Vienna transfers the renewal projects to private companies as well as limit-profit associations which have specialised in these complex renewal processes.

As stated above in the sector of limited-profit housing there is a special rent component which has to be invested in the object. Consequently, the problem of lacking financial means is not the main problem for housing associations to deal with. Of course there are very old buildings or buildings with special technical problems where this is not quite true, but in general it is more a question of the organisational capacity as well as of developing an appropriate scheme for renovation and modernization measures. More market-oriented strategies of maintenance and renewal have gained importance within the last years. As the aspiration level of tenants increased, the market position of some of the (older) dwellings is endangered. As a consequence some of the associations not only waited for upgrading the older dwellings when a tenant left or died but developed middle-term programmes for renewal; within these programmes they also try to amalgamate smaller dwellings, for which it may be necessary to find another accommodation for the sitting tenants. In other cases, the associations often carry out the improvements at turnover. In recent years even the demolition of existing older buildings and replacing them is being considered as an alternative. There are very few examples yet but it seems that in the future this could occur more often.

An increasing share of the housing associations' renewal activities is aimed at the reduction of energy consumption. The results of the housing survey of 1991 showed that the share of buildings/dwellings which had undergone measures of heat insulation within the last decade were the highest among limited-profit-associations. Similar results are found in the first incoming results of the survey that has been held in 2001. This indicates that at least the financial means and possibly also the sense of responsibility and know-how are more prominent in the housing associations sector than in other housing sectors. Of course one could also argue that the quality in other sectors might be higher, but that is doubtful at least as far as the old private rental stock and single-family homes are concerned.

3.3.5 Decision-Making Process: Considerations, Methods and Strategies

In former times, when housing was scarce, the only condition for landlords to invest was the availability of public grants. Nowadays, other aspects have become of similar importance and the decision-making process has become more complex. Lettability of dwellings is not any longer guaranteed and the choice of tenure, which has for long resulted in a clear-cut distinction between rent or sale, is more complex now rent with the option to buy has to be considered as well. Furthermore, social landlords in Austria have to deal with the problem of 'self-competition' between new developments and the existing stock, since it might happen that rents in newly built dwellings are (much) below the level of that in older buildings due the dynamic rent-pattern that is a result of the subsidisation system. At the same time, public grants are in most cases still a *conditio sine qua non* as rent levels due to the lack of own funds would be much higher since rents below the cost-coverage level are not accepted in the limited-profit system. In addition, we can identify three different types of factors within the decision making process of social landlords:

- cost factors (building land, construction, financing, management);
- demand and lettability in general (location of the project; technical and social infrastructure in the neighbourhood);
- competition, prestige, marketing.

While the first two types of factors prevent associations from making high-risk investments the latter one often leads to decisions bearing more risks.

Within the specific area of asset management, Austrian landlords (especially the municipalities) have to deal with the need for renewal of buildings that are often inhabited by sitting tenants who enjoy low rents. There are many factors and problems which have to be taken into consideration:

- the technical quality of the building;
- the market position (lettability) of the dwellings
- the available financial means for renewal, including subsidies and the level of financial reserves earmarked for renewal of the building;
- the acceptance of the activities and possible rent increases by the tenants;
- the planning and management capacities of the landlord's organisation.

The considerations and strategies of limited-profit housing associations focus on optimising the process of maintenance and renewal of the existing rental stock and fitting these activities into a feasible middle-term planning. And, although the sale of rental dwellings plays a minor role in the asset management of Austrian landlords, the transfer of the management of buildings to other associations—or the other way around, to take the management over—has become one of their more prominent strategies.

Since the limited-profit rental system does not leave much room for changing the existing rent levels, optimisation of the income-cost ratio has to be done mainly by reducing costs. The need for strategic cost-reduction has increased because of the historical peak in new construction in the past decade, which has placed the market position of some of the existing housing stock under threat. Consequently, housing associations have begun to analyse the development of the turnover rates. Furthermore, the associations analyse the cost-income-ratio of their estates to identify those estates in need of measures to improve their market position. The process could be, for example:

- defining the methods of monitoring; deciding which factors have to be surveyed—only 'hard' data, mainly cost factors, or also 'soft' data such as the working-load for the clerks that are responsible for the estate—and to develop a method of measuring and weighing these factors;
- deciding if all estates are to be included in that monitoring process or only those for which are assumed to generate excessive costs or problems of lettability;
- carrying out the survey and identifying the estates with a threathened market position and/or a poor cost-income ratio;
- deciding on the strategy to follow for these estates; landlords could, for example, choose between transferring the management to another association just to avoid excessive costs or trying to improve the cost-income ratio by renewal and/of financing strategies.

Within their analyses, associations do not only focus on the current performance of their own assets but also on the development of regional markets. These 'market-analyses' are carried out to be able to intervene into the rental costs of the existing stock in time to prevent 'house jumping' (either replacing loans by other ones or by own funds). In this respect it is also interesting to mention the changed role and activities of the Austrian Federation of Limited-Profit Housing Associations, which has recently begun to support these new strategies. Since a few years a system of benchmarking is in use in which data from the individual associations is collected into a model which can be used by the associations to compare their performance. Initially, only financial benchmarks for the overall economic performance were taken into account (at company-level), but now it is the aim to collect data about individual estates as well in order to provide better information to support the asset management.

3.4 CONCLUSION

The Austrian social rented stock is managed by municipalities and limited-profit housing companies. The majority of social rented dwellings is now in the hand of limited-profit associate, since the municipalities have largely withdrawn themselves from new housing production. In the future municipalities might even withdraw themselves further by transferring the management of their stock to housing associations, although this is not certain at this moment. In recent years, the market context for the social landlords has changed and this has influenced the landlords' management as well. After a period of an intense building boom and renewal activities, housing conditions have improved and quantitative and qualitative scarcity has decreased significantly. As a result, social landlords have to put more effort into the letting of some of their dwellings. Nevertheless, there is still much of public and political influence in social housing, among others through the housing promotion scheme which is concentrated on direct building subsidies, through political ties of the municipalities' management and housing associations' owners and network and through the legislative framework, particularly concerning the housing associations' rent setting.

The asset management has become more complex in the last years. It is not only a question of getting public grants for a project anymore, but there are also other factors which have to be taken into consideration. Now, the lettability has to be taken into account and the quality, price and location of new and existing dwellings have to be attuned better to the preferences of the households. As a consequence, housing associations have begun to use more sophisticated models of monitoring the income-cost ratios, lettability and market situation. In general, the asset management in the Austrian social rented sector, and that of the Austrian limited-profit-association in particular, is moving towards a more professional, independent and strategic approach. Of course, this does not count for all landlords to the same extent and landlords have to keep operating within the restrictions of the limited-profit system and the housing promotion scheme, but there is clearly a new orientation among the social landlords.

REFERENCES

Bauer, E. (1993a) Die öffentliche Wohnbauförderung in Österreich von den Anfängen bis 1960; in: Schulz, G. (ed.); Wohnungspolitik im Sozialstaat, Düsseldorf, Droste.

Bauer, E. (1993b) Wohnbaufinanzierung mit Zukunft; Wien, Österreichischer Verband gemeinnütziger Bauvereinigungen.

Bauer, E. (1994) Betriebswirtschaftliche Aspekte der Wohnbaufinanzierung; in: Korinek/Nowotny (1994).

Bauer, E. (2001) Gemeinnützigkeit im Wohnungswesen, Wien, ÖGB.

Blaas, W. et al (1991) Mehr Markt oder mehr Staat im Wohnungswesen, Wien, Böhlau.

Blaas, W., B. Brezina (1994) Zur politischen Ökonomie der Wohnungsgemeinnützigkeit; in: Korinek/Nowotny (1994).

CIRIEC (ed) (2003) Housing Statistics in the European Union 2002, Liège, http://mrw.wallonie.be/dgatpl/HousingStats.

Czasny, K. und P. Moser (1999) Einsatz und Gesamtwirkung der Wohnbauförderungsmittel, Fankfurt/Main, Peter Lang Verlag.

Donner, C. (1995) Das Ende der Wohnbauförderung; Wien, Selbstverlag.

Donner, C. (2000) Wohnungspolitiken in der Europäischen Union, Wien, Selbstverlag.

ECE/UN (ed.) (various volumes), Bulletin of Housing and Building Statistics, New York and Geneva, UN.

Eckhardt, J. (2001) Branchenanalyse und strategische Erfolgsfaktoren des Gemeinnützigen Wohnbaus in Ostösterreich, Diplomarbeit Wirtschaftsuniversität Wien.

Fassmann, H. et al. (1992) Haushaltsentwicklung, Wohnbau und Wohnungsbedarf in Österreich 1961–2031, Wien, ÖROK.

Fassmann, H., R. Münz (1998) Haushaltsentwicklung und Wohnungsbedarf in Österreich 1996–2021, Wien, ÖROK.

Holoubek, M. (1994) Die Geschäftskreisregelung als Kernstück des WGG, in Korinek/Novotny (1994).

Keinert, H. (1991) Grundfragen des zivilen Wohnungsgemeinnützigkeitsrechts; Wien/New York, Springer.

Kemmetmüller, W., M Schmidt (eds.) (1998) Genossenschaftliche Kooperationspraxis, Wien, Überreuter.

Korinek K. und E. Nowotny (eds.) (1994), Handbuch der gemeinnützigen Wohnungswirtschaft, Wien, Orac.

Ludl, H. (1998) Gemeinnützige Wohnbaugenossenschaften, in: Kemmetmüller/Schmidt (1998).

Lugger, K. (1994) Geschichte und Entwicklung der gemeinnützigen Bauvereinigungen in Österreich, in: Korinek/Nowotny (1994).
Matznetter, W. (1991) Wohnbauträger zwischen Staat und Markt, Frankfurt, Campus.
Matznetter, W. (2002) Social housing policy in a conservative welfare state: Austria as an example, Urban Studies, 39, 2, pp. 265–282.
Österreichischer Verband gemeinnütziger Bauvereinigungen (ed.) (various volumes) Die gemeinnützige Wohnungswirtschaft in Zahlen (Statistik des Österreichischen Verbandes gemeinnütziger Bauvereinigungen), Wien, Österreichischer Verband gemeinnütziger Bauvereinigungen.
Patera, M. (ed.) (1986) Handbuch des österreichischen Genossenschaftswesens, Wien, Orac.
Patera, M. (ed.) (1986) Perspektiven er Genossenschaftsrevision, Wien, Orac.
Scherz, B. (1994) Das dichte Netz von Aufsicht und Kontrolle, in: Korinek/Nowotny (1994).

4 BELGIUM-FLANDERS

SIEN WINTERS

4.1 SOCIAL LANDLORDS IN FLANDERS

Since 1980 social housing policy in Belgium has been entirely the responsibility of the three regions: Flanders, Wallonia and the Brussels Capital Region. The process of regionalisation was a gradual process, which took a long time. The law of 1 August 1974 regarding the provisional formation of regions stipulated that housing was a regional matter. This meant that, from 1975 onwards, the credits written into the national budget were divided among the three regions. Since 1980 the regional governments determine the credits for housing policy. In 1984, the Brussels Regional Housing Corporation and the Wallonia Regional Housing Corporation were created. The Flemish Housing Corporation ("Vlaamse Huisvestingsmaatschappij") was founded in 1988. The regional housing corporations started their activities in 1990–1991 (Mertens & Pacolet, 1998).

The legal basis for the housing policy in the three regions can be found in the former National Housing Code ("Huisvestingscode"), dating from 1970, which was intended to coordinate the legislation in force at the time. The Housing Code has remained to act as the legislative basis for the regional housing policy in the three regions during and even after the process of regionalisation. In the 1990s, the three regions each started to develop their own housing code. This resulted in the Brussels Housing Code ("Ordonnantie 9 September 1993"), the Flemisch Housing Code ("Vlaamse Wooncode", 15 July 1997), the Wallonia Housing Code ("Code wallon du Logement", 29 October 1998). So there was a common basis for housing policy in the three regions for a long time. And even since then, the major principles of social housing have evolved in similar ways in each of the regions (Hubeau, 2002). Although the legal context in which social landlords have to operate does not differ that much from region to region, this does not mean that asset management practices are also similar. For example, management contracts between the regional housing corporations and the housing associations were introduced in the Brussels Capital Region in 1994 and in the Wallonia Region in 1997 (Mertens & Pacolet, 1998). In Flanders, several sections of the Housing Code refer to the introduction of management contracts, but until today this has not been put

Vincent Gruis and Nico Nieboer (eds.),
Asset Management in the Social Rented Sector, 61–77.

Table 4.1 Division of the housing stock by tenure in Flanders

	Social rented	Private rented	Owner-occupied
Percentage of total housing stock	5.5%	22%	72.5%

Source: Estimations based on National Institute for Statistics, the socio-economic survey 2001 VHM, Jaarverslag 2002.

into practice. However, information or research on asset management practices for the three Belgian regions is not available. In our research we have confined ourselves to asset management practices in Flanders, the largest Belgian region.

The Flemish Housing Code ("Vlaamse Wooncode"—decree of 15 July 1997) starts with the basic principle of the right to decent housing. The housing code defines the objectives of Flemish housing policy, defines the tools and determines which players implement this policy. The social housing associations and their umbrella organisation, the Flemish Housing Association ("Vlaamse Huisvestingsmaatschappij"—VHM), are the preferred partners for the implementation of social housing. The social housing associations are private associations whose largest shareholders are the municipal councils. Their boards of directors consists mainly of municipal representatives.

Belgians are often said to be born with a "brick in the belly". Recent results of the latest socio-economic survey 2001 shows that 72.6% of the housing stock in Flanders is occupied by the owners of these houses. The social rented sector's share of total housing property in Flanders is very low. At the end of 2001, Flanders had 130,962 social rented dwellings. This is about 5.5% of the total housing stock. Slightly over half of these social homes are single-family dwellings, the rest are apartments (VHM, 2003)

At the end of 2001, 99 housing associations were active in the social rented sector in Flanders. The minority of these associations operates only in one municipality. The spheres of activity of the remaining associations cover several municipalities. Of the 308 municipalities in Flanders, 302 have joined at least one housing association. In 6 municipalities there is no housing association active. In most regions there is no overlap between the spheres of activity of the different associations. As a result, competition is not an issue within the sector. Only in Gent and Antwerp, the two largest Flemish cities, there are several associations active on the same territory.

The average housing stock of a housing association is 1,309 dwellings. The diversity between the associations, however, is very high. The smallest associations have 300–400 dwellings. The largest company has 5,730 dwellings.

Table 4.2 Some characteristics of housing associations in Flanders

Characteristic	Number/percentage
Number of dwellings	130,962
Number of landlords	99
Average dwellings per landlord	1,309
% of total housing stock	5.8%
% of social rented stock	Not available

Source: VHM, Annual Report 2001.

Another 5 have over 3,000 dwellings. The vast majority therefore has a stock of between 300 and 3,000 dwellings.

In addition to the 99 housing associations active in the social rented sector, 19 housing associations operate in the owner-occupied sector. Their main activities are building and selling houses to middle-class households (income limits are higher than for the social rented sector) and supplying subsidised mortgage loans of the VHM. Some housing associations in the social rented sector also build houses for the owner-occupied sector and thus combine both activities. A discussion is now going on concerning co-operation and merges between associations of both sectors.

Access to the social rented housing sector in Flanders is limited to low-income households (except from some projects for specific target groups). Based on current income criteria, half of the Flemish households are eligible for renting a social dwelling.

4.2 THE NATIONAL (FLEMISH) CONTEXT FOR ASSET MANAGEMENT

The asset management pursued by the housing associations in Flanders is formulated within a strict regulatory framework, based partly on Decrees issued by the Flemish Government or Ministerial Decisions and partly on the standards imposed by the VHM. Within this standardising framework, the housing associations have a moderate degree of autonomy and therefore limited room for pursuing their own forms of asset management.

4.2.1 Regulations

General administrative regulations

Alongside the VHM, the social housing associations are the preferred partners for implementing social housing policy. They have to ensure that social housing projects are fulfilled. Their tasks are described in the Flemish Housing Code as "the provision of social dwellings, revaluation of the housing stock and pursuit of a social land and buildings policy". In order to be able to claim government financing, the housing associations have to be recognised by the Flemish Government.

Allocations

The rental of social dwellings is regulated in the Social Rent Decree ("Sociaal Huurbesluit") Flemish Government decree of 20 October 2000). This defines eligibility conditions and allocation rules, as well as calculating rents.

To be eligible for a social rented dwelling, maximum income limits apply, depending on the family situation and the family may not own a home of fully enjoy the usufruct of property of the home. Prospective tenants should place their names on a register. The dwellings are allocated in chronological order of registration, taking into account a number of priority rules. Preference is given, for example, to people with a handicap if the available dwelling is suitable, to certain mobile home dwellers and to families leaving a home which has been declared uninhabitable, in cases of repossession.

When allocating homes, the housing association always has to bear in mind the rational occupation of the dwelling and the quality of life in the area. In order to be able to allocate correctly, the housing association must have internal allocation rules.

In the near future, the Social Rental Decree will probably be revised in order to allow housing associations more flexibility in their allocations and to achieve a better social mix.

Rent setting

The Social Rent Decree ("Sociaal Huurbesluit") defines the formula which has to be used to calculate the rent. This formula does not result in an exact rent, but defines a margin within which the company can fix the rent. Rent increases always have to be approved by the VHM. The main factors in the calculation formula are the updated cost price of the dwelling and the income and family situation of the tenant. The calculation is made as follows[1]: In an initial step, the company decides the basic rent for the dwelling, which is to be between 3% and 9% of the updated cost price of the dwelling[2]. The cost price includes all original building costs, as well as the

[1] We abstract some exceptional measures.

[2] The updating coefficient is determined by the government.

costs of improvement and renovation work. Secondly, the basic rent is adapted to the income and family situation. For this, the income coefficient (ic) is calculated:

$$ic = \frac{I + 1{,}576\ euro}{16{,}544\ euro}$$

where I income is reduced by € 993 for each of the first two dependents.

Larger families is granted a deduction from the basic rent. The deduction is 20% for 3 dependents, 30% for 4 dependents, 40% for 5 dependents and 50% for 6 dependents. Finally, maximum and minimum limits also apply. The monthly rent payable may not be higher than 1/60 of the income and not higher than the normal rent value (this is the rent for a comparable dwelling in the vicinity on the private market). The lower rent limit is half of the basic rent.

In addition to the rent, service costs and charges can be imposed, such as heating, gas, water and electricity, maintenance of communal areas of an apartment building, etc. Rents are recalculated on 1 January each year.

Sales

Social rented dwellings can be sold to the sitting tenants, but every sale of a social rented dwelling is subject to prior authorisation from the VHM. One condition is that rental property must remain available which, in terms of type, size and composition, is sufficient to satisfy the right to dignified housing with respect to the most needy families and single people. The quality of life and the statutory objectives of the housing association may not be jeopardised by the sale of social rented dwellings. The housing associations can judge this for themselves and therefore have a fairly wide degree autonomy in deciding whether or not to sell dwellings.

The "Transfer Decree" ("Overdrachtenbesluit") stipulates that the sale price must in any event be at least equal to the market price (using an official estimate), or to the updated cost price if the latter is higher. The sale price also has to be approved by the VHM.

The deed of sale (VHM model) contains obligations covering a 20-year period, relating to the personal occupation, disposal or transfer of a property right. In the event of default, the company can repurchase and, barring justified exceptional circumstances, payment can be imposed consisting of the non-written off part of the government investment.

The housing association must re-invest the entire proceeds from a sold social rented dwelling, within 5 years, in:

- building new social rented dwellings;
- in purchasing buildings and renovating or adapting them into social rented dwellings;
- or in land acquisition or renovation of its own property.

One of the present government's policies is to assign the sitting tenants the right of emption and to encourage the sale of social rented dwellings and to support this with financial tools. Government decisions on this subject are expected in the near future.

Maintenance and renewal

The VHM makes a distinction between maintenance and renovation. Work is covered by the heading "renovation" if:

- it involves an improvement in comfort and/or an improvement which leads to demonstrably lower consumption costs for the user;
- alterations with a layout change which implies an obvious functional improvement to the dwelling;
- work intended to change the number of dwellings within a building or to respond to changing needs (e.g. converting 2 houses into 3 apartments);
- Work carried out to dwellings or housing complexes which are under 15 years old is not regarded as renovation. The associations are obliged to reserve 10% of their rental incomes for future renovation and maintenance. There is no other regulation concerning quality and maintenance of social houses and there is no minimum standard for social rented houses.

4.2.2 Supervision

The VHM is responsible for supervising the housing associations. Based on project proposals from the associations, the VHM draws up investment programmes. In addition, the VHM fulfils a banking function. It raises financial resources from the capital market and, with subsidies from the Flemish government, provides attractive loans to the housing associations for building and renovating social dwellings. The housing associations are obliged to transfer their financial reserves to the VHM.

The VHM ensures that the social housing associations carry out their tasks and that the laws, decrees, regulations and statutes are observed. This supervision can take place via prior authorisation, approval or ratification of a decision made. The supervision relates to the social, administrative and accounting obligations. This means, for example, that the associations always have to have authorisation from the VHM in order to undertake property transactions, for external investment financing, for the award of building work to contractors, or for making their property available for purposes other than social housing. They have a duty to report financial statements and accounts, the minutes of general meetings and the appointment of directors.

Technical supervision on investments is highly developed. For building work that is worth more than €21,973, the associations are obliged to have the preliminary draft approved by the VHM. For building and renovation projects, the VHM uses general design criteria, price and surface area standards, performance regulations and recommendations.

In addition to supervision by officials of the VHM, each association has a VHM supervisory director with supervisory powers. The supervisory director can participate in all meetings of the management bodies of the association and has access to all documents. He has the power to suspend implementation of decisions which he regards as incompatible with the legislation and regulations.

The general feeling in the sector is that the VHM exercises too much supervision over details and not enough over final results. The associations perceive supervision by the VHM as meddling in a number of situations and rely on their status as private associations in order to make their own decisions. The present forms of supervision are also labour-intensive and therefore expensive. On the other hand, the VHM has few sanctioning options if infringements are confirmed. The only possibilities are currently for the VHM to act in the place of the directors of the company and to withdraw its recognition. This possibility is only used in extremely exceptional circumstances.

Partly in order to respond to these problems, the Flemish Housing Code includes the provision that management agreements should be reached between the VHM and the housing associations. These should make arrangements concerning the objectives and the results to be achieved by the company, in exchange for a number of commitments by the Flemish Government, including that the company could have an investment envelope at its disposal for several years. The associations could thus gain more autonomy. The role of the VHM would be correspondingly less supervisory and more supportive. Moreover, because of the difficulty of reconciling its roles as supervisor and adviser, the supervisory tasks of the VHM will be transferred to the administration of the Flemish Community. However, more than 5 years since the introduction of the Flemish Housing Code, no management agreements have been reached, although much work has been done in this direction. One of the important elements of this is what is now referred to as "mutual positioning" ("onderlinge positionering"). This is a system of benchmarking, in which the associations have the opportunity to compare their company's results with those of comparable associations from the same peer group. This is therefore a form of performance assessment which can later form the basis for arrangements within management agreements. Consultation is currently underway between the VHM and the associations in order to reach "working arrangements" based on this information, which in fact boils down to a soft form of management agreements (see also section 4.3.5).

4.2.3 Financial Support

In order to implement their tasks, the social housing associations can make use of the following sources of financing:

- project subsidies from the Flemish Region;
- loans from the VHM, which raises loans from the capital market for this purpose and receives an investment grant from the Region;
- financial reserves from the past, with the consent of the VHM;
- loans from third parties, with authorisation and approval from the VHM.

Investment in new construction and renovation is subsidised by the Flemish Region. Two important financing mechanisms exist: the 'investment programme' and the 'project subsidies'. Between these two financing systems the degree of subsidy is different. Also the administrative application procedures are differing: the VHM is responsible for drawing up the 'investment programmes' once a year, the Flemish administration decides about the project subsidies. This dual system makes it difficult to draw long term investment programmes based on a systematic analysis of needs, which are importantly differing between regions and between target groups. Complementary to these investment subsidies, housing associations can apply for subsidies financing investment in infrastructure (roads, parks, ...), for acquisition of land or property, for projects for specific target groups, e.g. for handicapped persons.

The level of the subsidies is depending on certain cost elements, but not on the income of the housing associations generated by these investments. In particular, this future income (rents) are highly related to the household income of future residents. Since, in general, incomes of the tenants are decreasing, the financial position of many housing associations is under pressure. Especially in the cities incomes of tenants are low and housing associations are now facing severe financial problems, causing an important obstacle to new investments and renovations.

Incorporated in the financing mechanism there are some forms of solidarity between 'poor' and 'rich' housing associations. Within the "investment programme" housing associations with a low "ic" (see 4.2.1) enjoy a slightly more favourable loan repayment scheme. However, the advantages are very limited. Recently, these associations with a low "ic" were granted with a special subsidy aimed at covering the gap between income (rents) and expenditures. The result of all this is a complex financing mechanism which lacks transparency.

4.2.4 Housing Market

There are no figures available concerning quantitative shortages on the Flemish housing market, but rising prices suggest that shortages emerge. For renting a social dwelling there are long waiting lists. At this moment 60,000 households have subscribed for a social dwelling and are still waiting. The average waiting time mounts up to 2 to 3 years, for some dwellings it can be 6 years or more.

The latest figures on housing quality show that the Flemish housing stock is in a rather bad condition. Compared to other European countries Flanders is not performing well (Winters, 2001, based on EU Housing Statistics). This is partly the result of limited grants for housing renewal. Compared to households on the private rental market, low-income households in the social rented sector have a high quality house rented for a low price on average. Tenants are in the social rented sector are fairly satisfied (Pannecoucke, 2002), while on the private rental market, affordability problems, related to substandard housing quality are increasing.

4.3 ASSET MANAGEMENT PRACTICE

A systematic description of the choices relating to asset management, made by the associations within the above regulatory framework, is not available for Flanders. From our knowledge of the sphere of activity and based on a limited number of interviews with representatives of the housing associations and VHM officials, we believe we can assert that practices in this respect differ greatly, depending on factors such as the size of the company, the training and experience of the managers and the expertise of directors. In general, it can be said that the vast majority of the housing associations pursue an *ad hoc* policy to a significant extent: decisions about investments, the setting of rents and the sale of dwellings are not based on long-term plans at technical, financial or social level.

4.3.1 Rent Setting

The housing associations have a certain degree of autonomy in setting rents, especially via the establishment of the basic rent within the margins of 3% to 9% of the cost of the dwelling. The main factor in this decision is the general financial situation of the association. In setting the rent for a group of dwellings, only a very tenuous link is made with the expected expenditure flow for the project. If this link is made at all, short-term considerations, often based on expenses in the coming year, are dominant. If, for example, at the beginning of the year an increase in costs for a particular area or apartment block is expected as a result of planned maintenance work and these costs are not covered by the rent, this will be taken as a basis for the decision to increase the basic rent. One of the reasons why the expenditure flow and the rent level are not related is that no tradition of cooperation exists within the housing association between the renting and accounting departments.[3]

Some associations have a strong tendency, based on social considerations, to keep rents as low as possible. This attitude is wide-spread among the directors, the majority of whom have political mandates from a municipal council. Increasing rents is a politically sensitive issue. At the same time, these associations seem unaware of the consequences of this for the profitability of the association and the long-term investment possibilities. According to a privileged witness from the sector, the majority of the associations actually take not increasing the basic rent as an explicit point of departure; decisions about maintenance and renovation are subordinate to this principle. In other words, this work is put off infinitely.

4.3.2 Allocations

As described above, the housing associations have only a limited margin within the legal framework for pursuing their own allocation policies. Some associations make creative use of any available margin for pursuing their own allocation policies, for instance for renting to specific target groups such as the elderly and handicapped. By contrast, other associations hide behind the legislation so as not to pursue a target-group policy. Within the sector, many are currently calling for a more flexible allocation policy, in particular in order to respond to quality-of-life problems and to achieve a better social mix.

Residents of which the income is too high to be eligible for a social rented dwelling pay a surplus on normal rents when they prefer to stay in the house. By this way households are stimulated to move, leaving the house for households with stronger social needs. However, there is a political agreement between the coalition partners of the Flemish government to evict tenants with too high incomes (ic > 2). In December 2002 many households received a letter

[3] One might expect a more professional financial management and therefore a better cooperation between the rental and the accounting departments in the larger housing associations. However, financial management at this time is not that developed, even not in the larger associations. Some directors of housing associations reported that smaller housing associations due to the limited size often have a better overview of their financial situation than the larger associations, where the partitions between the various departments are thicker.

that for this reason they had to leave their house within two years. Strong reactions from tenants and housing associations, however, have resulted in postponement of this agreement.

4.3.3 Sales

In 2002 157 dwellings were sold in Flanders to sitting tenants. During the period from 1992 to 2002 a total of 3,050 dwellings were sold, so the number of sales is very low. The associations decide themselves whether to sell the dwelling at the request of a tenant. Information about the motives of associations to sell is fragmented and therefore inadequate, but it appears that improvement of the financial situation always plays a role. The legal condition for sale (that the remaining social rented property must be sufficient in terms of nature, scope and composition to satisfy the right to dignified housing for the most needy families and single people) is almost never met, given the limited social housing stock in Flanders.

4.3.4 Maintenance and Renewal

The planning of maintenance and renovation does not take place systematically or in the form of long-term planning. The VHM requires that 10% of annual rental income is reserved in a financial provision for maintenance and renovation. Many of the associations do not abide by this requirement, in particular because their annual accounts, even with no provision, are already close in the red. The VHM has no resources to take action against these associations (see section 4.2.2).

Some larger associations try to approach their maintenance planning in a somewhat more systematic way. For example, they use lists of roof coverings which have to be replaced every 10 years, paintwork which has to be done every 5 years, etc. Some associations have sought external advice for drawing up these plans. To improve this situation, the VHM has begun to make an inventory of the state of the social rented housing property (see the box in the following section).

4.3.5 Decision-Making Process: Considerations, Methods and Strategies

In Flanders, the housing associations in the social rented sector pursue strategic asset management only to a very limited extent. Decisions about managing social rented dwellings are not taken on the basis of systematic analysis of the internal and external environment or on the basis of long-term objectives at financial, technical and social level.[4]

[4] This is the case for asset management decisions, as well as for new developments. The planning of projects by associations is not preceded by any systematic exploration of the market or of

Housing associations do not follow a cyclical planning process starting from determination of objectives and ending up in evaluation by several stakeholders, aiming to reformulate aims and objectives. Performances are not systematically and periodically evaluated and decisions on asset management mostly are taken on ad hoc basis.

The housing associations also operate only a very limited form of financial planning. The influence of investments on the financial performance of the association in the longer term (profit and loss, expected development of liquid assets, etc.) is not systematically examined. The associations do not possess the tools to make this evaluation.

One explanation for the lack of any long-term planning is the serious complexity of the financing system. In the past, the financing and subsidy mechanism was changed several times and new systems were added to the existing one. These financing systems have various redemption mechanisms and different levels of subsidy and are dealt with via various administrative procedures. The associations have outstanding loans within each of these systems. This leads to a lack of transparency and does not encourage the associations to manage efficiently. In conjunction with this, some managers of associations give the explanation that management of their financial resources does not take place within their own associations, but by the VHM, which does not encourage them to take responsibility. Moreover, they would have insufficient insight into the financial flows of their association. In response to this, the VHM is working on transparent communication concerning financial transactions and on the development of support tools for financial planning (see below).

The VHM wants to support the housing associations as much as possible in pursuing good management in general and asset management in particular. To this end, a number of tools are currently being developed which are being offered to the associations. The three main tools are 'mutual positioning', inventory of the state of the property and financial planning (see frame). The aim of each of these tools is to offer the associations a better reference framework for self-evaluation and for evaluating decisions, so that they can take the best decisions within their present (limited) autonomy and, in the long term, be in a position to acquire greater independence. We explain these tools below.

---

the needs of the target groups. Projects are rather the result of local opportunities (e.g. the sale of a large building plot, the demolition of a school, etc.) and of the availability of financial resources. If an opportunity presents itself, it is taken. At the same time, there are many obstacles. Several associations—particularly those active in the cities—are battling against serious financial shortages. Another, more common problem is that municipal councils do not want any more social housing on their land because they fear that this poses a risk to the quality of life and to municipal finances.

Mutual Positioning

"Mutual position" is a system which allows the associations to situate/position themselves in the sector and to carry out self-inspection, as well as providing management information for the VHM. The input and processing of the data works via the "housing network"; this is a secure electronic communication network linking all associations to the VHM[5]. In order to be able to interpret the management information in a meaningful way, the method chosen was one in which the social housing associations are divided into peer groups, based on the nature of their activities, the size of the company and the area in which they operate (in large towns or not). Once the associations enter the information about their own operations into the system, they can position themselves with respect to the Flemish average and with respect to the "peer group" average. The following information forms part of mutual positioning:

- property: number of dwellings rented, vacancies, etc.;
- sale of rented dwellings since 1997;
- background to the income coefficient (see price setting);
- background to rent arrears;
- building activities: new construction, renovation, maintenance (including costs for renovation and maintenance);
- financial indicators: accounting value of property, liquid assets, operating income, etc.;
- environmental factors in the sphere of activity: housing market indicators, demography, etc.;
- management aspects of the company: expenses reimbursement, entertainment expenses, etc.

The system was tested in 2002 and is currently being adjusted. At the same time, the VHM and the associations are now trying to reach a model for working arrangements, based on this system.

Inventory of the State of the Property

In 1999 the VHM began an intensive project to make an inventory of the state of social rented housing property. The intention was to achieve a uniform way of describing the property for all associations, in the first

[5] Since the associations themselves can decide to participate in the "housing network", not all associations are connected to it yet.

instance so as to gain a global view at Flemish level of the renovation needs in the sector and to be able to provide the required funds in the budget and, secondly, to give the associations themselves a clear insight into the maintenance and renovation needs which they will face in the future.

In an initial step, architects and engineers from the VHM developed a detailed questionnaire which allows the technical quality of social dwellings to be described. Then, an information technology package was developed which can be integrated into the various software systems used by the various associations.[6] Since early 2000, 5 engineers have been sent out every day to the associations in order to gather data. For this purpose, use is made whenever possible of the information present at the administrative head offices, which is then supplemented during on-site visits. The first round of information gathering is therefore being undertaken by the VHM. Subsequently, it is assumed that the associations will permanently update the data.

At technical level, a distinction is drawn between maintenance planning and renovation planning. In the first instance, the inventory examines, at district level, to what extent the various building components are due for maintenance. Periodic maintenance is linked to the state of the building components (period from one year up to the life of the building), with a price. The intention is that the company can make provision for this in its budget, based on a genuine plan.

In the second part, the functional condition is established per dwelling group, comprising the number of bedrooms and the type of dwelling. This forms the basis for the renovation plan. In principle, increasing age is associated with increasing priority for renovation. Furthermore, "fields of action" are indicated per age category, in decreasing order of priority: safety (inspection of gas and electricity systems), insulation (wind insulation, roof insulation), comfort (checking the presence of a flushing toilet in the house, bathroom with sanitary hot water, possibly central heating). Furthermore, a number of requirements imposed by other bodies are also checked. A link also exists between maintenance and renovation activities because if renovation is carried out, maintenance work becomes superfluous.

---

[6] Computerisation in social housing has progressed during the 1990s, primarily at the initiative of the individual associations. Consequently, the associations work with different software suppliers, which makes it difficult for the VHM to provide uniform information technology applications.

Financial Planning

The VHM is currently fully occupied with developing a software tool which should allow the associations to draw up a long-term financial plan. The system is based on the IAS standard 7 concerning cash flows. Based on this software, the associations will be able to make forecasts about their treasury and profit and loss in the future. This will also allow them to estimate the financial implications of investments in new building and renovation and to use these as a basis for their decisions.

Stakeholders involvement

Stakeholders (e.g. municipalities) are (or at least should be) well informed about the results of the housing associations via their representatives in the Board of Directors and do are involved in daily decision taking of the housing association. One of the problems is that many housing associations are small scale and that the representatives of the local authorities lack knowledge and professionalism. Also a problem is that being member of a Board of Directors of a housing association is not a priority for local politicians. In public opinion social housing does not have a good image.

Tenants are not directly involved in the decision making process, but tenant participation hasgained importance during the 90s. This was partly a result from the urban policy during this decade, which was strongly directed towards deprived urban areas, among which many social housing neighbourhoods. Plans were drawn up to revitalise these neighbourhoods, in close cooperation between central actors (communities, housing associations, welfare organisations, etc.) and citizens. At the moment, the level of tenant participation still differs widely between housing associations and is strongly project-related.

4.4 CONCLUSION

In Flanders, the housing associations in the social rented sector pursue strategic asset management only to a very limited extent. Decisions about managing social rented dwellings are not taken on the basis of systematic analysis of the internal and external environment or on the basis of long-term objectives at financial, technical and social level.

Within the typology of approaches towards asset management, as described in the introduction of this book, the Flemish housing associations can be characterised as task-oriented. The decision-making process reflects an incremental approach, involving fragmented and small scale series of changes and the housing associations do not formulate (long-term) objectives for the development of the housing stock. However, in a certain way many housing associations can be characterised as proactive. The actively seek for investment opportunities stemming from the local housing market. On the other hand, financial and technical problems are often solved only after the problems emerge.

Following Kemeny (1995), the Flemish housing market is a dualist rental system, still strongly regulated and controlled by the (regional) government. This partly explains the lack of strategic asset management in Flanders. The market does not force the associations to follow sophisticated strategies. Particularly at the level of financing investments, allocating homes to tenants and setting rents, a strict regulatory framework exists. On the other hand, we observe that decisions that the associations can take themselves within this regulatory framework are not based on strategic considerations either. So, there are more explanations. First, social housing in Flanders is of relatively high quality and that the demand for social rented housing is particularly high. Thus, associations are not stimulated by market forces to adopt a more strategic approach. Second, the financing mechanism is extremely complex and not transparent. Third, the financial resources of the associations are managed by the VHM, which does not encourage the associations to take responsibility themselves. Fourth, insecurity concerning future income flows makes long-term financial planning more difficult. A fifth possible explanation is the relatively limited size of the housing associations, together with the lack of competition among associations. An increase in scale and professionalisation would probably contribute to more strategic asset management.

In recent years, the VHM has tried to bridge the current gap by offering tools to the housing associations to form the basis for better asset management. The intention of these tools is to encourage the associations to test their decisions against those of comparable associations in the sector and to be able to estimate the influence of these decisions on the quality and size of their property and on their financial situation in the long term. Within the current institutional framework, this is already a good option. However, the question is whether the entire institutional framework is not due for revision, so as to encourage the housing associations to pursue efficient and effective policies. Introducing some principles of market forces could contribute to this.

REFERENCES

Bloemen H. (1998) De Vlaamse wooncode, Vanden Broele, Brugge.

Hubeau B. (Ed.) (2002) Vijf jaar Vlaamse Wooncode: het woonbeleid (nog niet) op kruissnelheid?, Brugge, Die Keure.

Mertens M. & Pacolet J. (1998) Management contracts in social housing in Belgium. Three different routes and initial lessons, in: Pacolet J & Mertens M. (eds.) Management Contracts and Performance Indicators in the Social Housing Sector: Belgian and European Experiences, HIVA-KULeuven, Leuven, pp. 59–86.

Meulemans, B. en Willemé, P. (1999) Wonen Becijferd. Woonbehoeften in Vlaanderen, 1991–2010, in: De Decker, P. (ed.) Wonen Onderzocht 1995–1999. Een overzicht van de resultaten dan de recente onderzoeken, Brussel, Ministerie van de Vlaamse Gemeenschap, pp. 45–68.

Pannecoucke I. e.a. (2001) Profiel van de sociale huurder en subjectieve beleving van de realisaties van de sociale huisvesting, Antwerpen, UFSIA, Antwerpen.

Ministerie van de Vlaamse Gemeenschap (2002) VRIND; Vlaamse regionale indicatoren 2001, Brussel, Ministerie van de Vlaamse Gemeenschap.

Vlaamse Huisvestingsmaatschappij (2002) Jaarverslag 2001, Brussel, VHM.

Winters S. (2001) Lifetime housing in Europe: the rationale, the strategies, the policy instruments', in: Winters S. (ed.), Lifetime housing in Europe. proceedings of the European seminar and colloquium 'Living Tomorrow Starts Today', 4–5 December 2000, Leuven, HIVA-K.U.Leuven, pp. 5–30.

5 DENMARK

LARS A. ENGBERG

5.1 SOCIAL LANDLORDS IN DENMARK

The social housing sector in Denmark is owned and managed by private non-profit housing associations governed by democratically elected tenants. Despite the private legal status, the sector is subject to detailed public regulation. The key rationale in the sector is to provide universal access to adequate housing. To realise this objective, social housing rents have been regulated in order to make them stay within the means of persons with relative low incomes. In principle, access to social housing in Denmark is universal but demand exceeds supply and access is restricted by a system of waiting lists. 25 per cent of all housing units are reserved for persons with special housing needs, a scheme administered by local government.

The social housing stock accounts for about 20 per cent of the total housing stock, totalling about 495,000 social housing units (see Table 5.1). Most of these housing units are relatively new, as only 5 per cent have been built prior to 1940. Since 1980 the number of social housing dwellings has risen by 50 per cent. In the last two decades the relative share of social housing has risen compared with private rental housing, and in 1998 the number of dwellings in the social housing sector exceeded that of privately rented dwellings.

The Danish social housing stock is administered by about 700 social housing associations. The social housing association ("almen boligorganisation") tends to the primary management of the housing stock. Also, the housing association administers the allocation of flats and makes decisions to initiate new building projects, which however have to be sanctioned by local government.

In Denmark, there is a strong tradition of tenants' democracy and self-governance in social housing, and tenants' representatives in fact govern the social housing associations. Tenants who live in individual housing sections elect *section boards* ("afdelingsbestyrelser") responsible for daily management and financial governance of each section. However, the primary decision-making unit is the *tenants' assembly* ("beboermødet") where matters relating to section budgets, physical renewal of the estate and collective house

Vincent Gruis and Nico Nieboer (eds.),
Asset Management in the Social Rented Sector, 79–95.

Table 5.1 Division of housing stock by tenure in Denmark (1999)

	Social rented	Private rented	Owner-occupied
Percentage of total housing stock	20%	18%	58%

Source: Bolig- og Byministeriet 1998: Bygge- og boligpolitisk oversigt 1998–1999, p. 77.

rules ("husorden") regulating every-day behaviour in the estate are decided upon.

Each association is subject to the authority of a governing body, a *housing council* ("repræsentantskab") composed by a majority of tenants' representatives and the members of *the executive board of directors* ("selskabsbestyrelse") which is nominated by the council. As the governing body, the housing council has to adopt the annual budgets of the housing association, and it decides upon all major administrative and financial issues of the association, i.e. whether to sell real estate (non-housing units) or initiate major physical changes in the housing stock. Danish social housing associations are not entitled to engage in the construction of non-social housing units.

The housing council can delegate specific prerogatives to the level of section boards. The executive board of directors is responsible for the implementation of policies decided upon by the housing council, and the committee appoints a housing director in charge of daily management. The chairman or deputy chairman has to be a tenant, and tenants constitute the majority in the board of directors.

Prior to 1998 three types of housing associations existed, co-operative societies, self-governing associations and joint-stock companies differing somewhat in their organisational features, but with the Social Housing Act of 1997 these differences have been harmonised in line with the outline above.

Table 5.2 Some characteristics of social housing associations in Denmark

Characteristic	Number/Percentage
Number of dwellings	471,000
Number of landlords	about 700
Dwellings per landlord (average)	about 700
% of total housing stock	20%
% of social rented stock	100%

Source: Bolig- og Byministeriet 1998: Bygge- og boligpolitisk oversigt 1998–1999, p. 77.

5.2 NATIONAL CONTEXT FOR ASSET MANAGEMENT

The organisational dynamics of Danish social housing associations with respect to asset management should be understood in the institutional context of the social housing sector as such. The key features of the sector are these:

- the social housing sector is a non-profit sector based on the principle that rents and expenditures should balance;
- there is a tradition of tenants' democracy and self-governance;
- the social housing sector is subject to a very detailed and comprehensive legislative regulation.

Because of these features, the strategic scope for "asset management" in terms of market-oriented behaviour of social housing associations is very limited in the Danish case. Traditionally, social housing associations have supplemented the core task of housing provision with related business activities in areas of housing maintenance, repair works, care-taking etc., but the scope of these supplementary activities has been reduced through legislative changes. However, the current the Liberal-Conservative government is pushing for privatisation, and the threat or challenge of a possible sale of social housing is prompting reform initiatives from the sector itself. These initiatives primarily focus on rationalisation of housing administrations, individualisation of housing services and the possible increased financial involvement of tenants in terms of semi-ownership of individual dwellings.

5.2.1 Regulations

General administrative regulations

The primary laws regulating the social housing sector are the Social Housing Act of 1997 and the amendment of The Social Housing Act of 1998. Social housing is subject to detailed regulation in a number of secondary laws and regulations.

Rent setting

The rental policy in Danish social housing is based on the principle of cost-price renting. The costs of living in social housing should not exceed the direct building expenses minus public subsidies in combination with financial and administrative costs of running the estates. Profits that may accrue are earmarked for building and maintenance purposes. Each local housing section is in principal a financially independent unit, and no cross-subsidy or pooling of costs between housing sections is in principle possible. Because tenants decide upon the level of expenditures (in co-operation with the housing association) a policy of rent-stability is typically pursued with great zeal.

Allocations

As stipulated by law, the allocation of housing units in the social housing sector is regulated in a system of external and internal waiting lists. Everybody can subscribe to the external waiting list while the internal waiting list is only for tenants who already live in social housing apartments and who want to move on to what they consider a more desirable apartment. The internal waiting list has priority over the external list with the consequence that newcomers have limited access to apartments, in particular on the most attractive social housing estates. In some cases the internal list only applies to a single housing section, in others it applies to all housing units in one or several housing associations.

According to law, local authorities are entitled to dispose of every fourth vacant social housing apartment in a municipality. The rule makes sure that the municipality has accessible housing available for persons with a special housing need. If a municipality makes use of the scheme it has to cover costs associated with vacating the housing units and to put up a guarantee for the tenant's deposit.

Not all municipalities lay claim to this right, often a voluntary agreement between one or more of the larger housing estates and the municipality exists in which vacancies are allocated to the social administration of the municipality on an ad hoc basis. With the Social Housing Act of 1997 a pilot scheme was introduced allowing housing associations and municipalities to experiment with housing allocation rules. The idea behind the scheme is to allow for a more flexible allocation policy taking that takes into account local circumstances and promotes a more balanced composition of tenants (Boligog Byministeriet, 1998b). The scheme introduced the possibility that housing associations can abolish the waiting list system and reserve a number of apartments for specific target groups and that local government may dispose of more than 25 per cent of vacant dwellings.

Sales

Up till now, the selling of social rented dwellings is not allowed. However, the current Liberal-Conservative government wishes to liberalise the social housing sector and allow for sale of social housing units. The government has set up a working group that looks into the legal, financial and other ramifications of sale. The working group has issued a proposal that is met with opposition from social housing associations and local governments both. In the proposal it is suggested that each housing section can decide whether tenants are handed the right to buy their apartments. Flats are sold to the market prize, but tenants only pay 70 percent of the market prize when they buy the apartment and the remaining 30 percent if they choose to sell it. The housing section would receive a small financial bonus from the sale.

Maintenance and renewal

Housing associations are required to set up long-term maintenance plans for all housing sections, and tenants and housing professionals carry out annual on-site evaluations of the estates and decide upon a maintenance policy.

Tenants influence the budget of the housing section within defined parameters, for instance a minimum level of appropriations is required by law. A local fund covers maintenance and other running expenses, and rents are also channelled into the National Building Fund that assists the continuous maintenance and renewal of the sector.

5.2.2 Supervision

There is no national supervisory body in Denmark. Local government has played a key role in the running of social housing associations together with labour unions that have played a historical role in the development of social housing. The primary role of local government is administer the social housing quota and to supervise that social housing administrations comply with the comprehensive and detailed legislative framework regulating their activities. To perform this task the city council has to sanction a number of decisions relating to the financial and administrative operations of housing associations.

The city council has to make sure that the physical condition of housing estates is acceptable, and it sanctions the level of rents in new and existing housing estates and control that the principle of a balance between costs and expenditures is maintained. For this end it carries out an annual auditing of the housing association's accounts. If the housing association decides upon a change in statutes this change has to be approved by the city council.

Following changes introduced with the Social Housing Act of 1997 local government is no longer automatically represented in the governing body of housing associations. It is now up to the housing council to decide whether the municipality should nominate a representative to the board of directors.

5.2.3 Financial Support

Local government covers 7 per cent of the building expenses of new social housing estates and has to provide guarantees for loans covering new housing constructions and major repairs. Tenants pay 2 per cent and 91 per cent is covered by a loan from a private mortgage institute. Tenants' share of repayments amount to 4.3 per cent of annual capital expenses and the difference between this share and overall repayments is subsidised by the state. In 1998, these subsidies amounted to DKK 4.9 billion (€ 0,67 billion), three thirds financed by central government and one third by local government. When mortgages are paid back tenants continue to pay rents which are channelled into funds allowing for the continuous modernisation of the existing housing stock as well as the construction of new social housing units. Previously, a maximum ceiling on production costs existed stipulating that these should not exceed a maximum amount per square meter, in 1995 the recommended price ranged from 9,980 DKK/square metre (1,340 €/m^2) in Copenhagen to 8,400 DKK/square metre (1,128 €/m^2) in some rural areas in Jutland. However, as from 1 January 1998 this maximum ceiling is no longer in force.

Each social housing estate is treated as a financially independent unit. The social housing association is not responsible for the financial liabilities of housing sections and vice-versa. The social housing association has a *disposition fund* that functions as a financial buffer, allowing a certain re-allocation of money between housing sections. The fund collects obligatory financial contributions from housing sections and tenants payments from housing sections that have paid back their mortgages. The disposition fund subsidises housing sections with financial difficulties due to voids, major repairs etc., and it contributes towards the financing of new housing constructions.

Furthermore, housing associations have a working capital that cannot exceed a certain maximum ceiling defined by the city council, and profits (administrative fees, dividends) that may accrue are channelled back to the disposition fund and earmarked for building and maintenance purposes. The National Building Fund guarantees the housing associations administration of reserves and operating funds from the housing sections.

The National Building Fund

Tenants who live in housing sections built prior to 1970 pay an annual contribution to *The National Building Fund* ("Landsbyggefonden") as a part of their rent. Tenants' contributions amount to approximately DKK 620 million (€ 83 million) per year, and the Fund re-channels the funds back to the social housing sector through 3 separate distribution mechanisms:

- Each social housing association is entitled to 60 per cent of its contributions paid since 1980 in order to subsidise housing constructions, repairs and maintenance;
- 27.5 per cent of annual contributions are earmarked for a *modernisation fund* that subsidises shoddy construction work and environmental programmes. Only housing sections planned prior to 1 July 1986 have access to the modernisation fund;
- 12.5 of annual contributions are put into the *National Disposition Fund,* a fund that subsidises running expenses of less well-off housing sections.

The legislative framework regulating the financial aspect of the social housing sector has undergone a number of changes, shifting the main burden of subsidising social housing from the state to the sector itself. The amendment to the Social Housing Act of 1998 introduced a finance reform of the sector. The main purpose of the reform was to reduce state expenditures in connection to new housing constructions by gradually increasing self-financing in the sector in a long-term perspective (Lovforslag nr. L 46). Social housing has been financed with price regulated index bonds making the period of repayment dependant on the inflation rate. With the finance reform, bonds with a fixed 30-year repayment period were introduced with the effect that tenants' user payments increase after 20 years with a proportionate reduction in public expenditures.

With the amendment of The Social Housing Act of 1998 the fund will receive more resources in the future. After 1 January 2000 half of all tenants payments from existing housing sections in which mortgages have been paid back will be directed to the fund, and one third of payments from new housing units after tenants contributions have been frozen after 35 years of repayment. The reform also introduces a new *Housing Construction Fund* ("Nybyggerifonden") set up to collect rents from housing financed after 1 January 1999. The financial flow into the fund will gather momentum when mortgages will be repaid in 30 or 40 years.

In October 2002, the Parliament voted for a housing agreement (L 53) that draws from the growing resources of the National Building Fund DKK 2.7 billion (€ 0.36 billion) per year for building and renovation activities. A new housing renovation scheme has been set up and the fund takes over the state obligation to subsidise new social housing units in the period 2002–2004 as well as a number of additional financial obligations previously carried by the state.

The Urban Committee action programme

In 1993 the Social democratic coalition government established an Urban Committee as a response to a 'ghetto-debate' raised by a number of social democratic mayors from the Greater Copenhagen area. The Urban Committee initiated an action programme targeted at the social housing sector to combat the negative consequences of social and ethnic segregation. The key elements in the programme were:

- A re-financing scheme of DKK 10 billion (€ 1.3 billion) strengthening the market position of newly built social housing estates by lowering rents while financing a physical 'face-lift' and the initiation of social and cultural initiatives on the estates.
- A lump-sum of DKK 1.6 billion (€ 0.2 billion) covering the period 1994 to 1997 to support a social development strategy in deprived housing estates.

> The fund financed 8 innovative Action Models; formalised partnerships between municipalities, the social housing sector and other interested parties co-ordinating area-based social development strategies. In addition, 100 social workers were employed as community developers in various social housing estates (Vestergaard et al., 1997).

About 500 housing sections totalling 155,000 dwellings have received financial support from the Urban Committee in the general effort to strengthen the market position of the social housing sector. About half of these were situated in the Greater Copenhagen Area and one fourth in the largest provincial cities. All housing sections were struggling with high vacation levels, a concentration of social problems, physical decay and stigmatisation.

The re-financing scheme resulted in revenues of DKK 6 billion (€ 0.8 billion) equal to an annual rent reduction of DKK 430 million (€ 58 million) for a 30-year time-span (Pedersen, 1998). Housing sections subject to the re-financing scheme set up new loans in total sizing the revenue of DKK 6 billion (€ 0.8 billion) in order to finance physical refurbishment, and the overall outcome of the process was a balance between revenues derived from mortgage re-financing and the physical modernisation of the estates. The relative distribution of rent reductions and physical improvements differed between estates and in average the outcome of rent reductions varied from DKK 48 (€ 6,40) per square metre to DKK 61 (€ 8,20) per square metre (ibid).

5.2.4 Housing Market

On the Danish housing market, costs and rents are regulated in the rental, social and co-operative housing sectors. As a consequence, housing costs in these three sectors are lower than in the market regulated private sector. Because of this, demand exceeds supply in the regulated sectors; since 1994 there has only been about 200 voids annually out of the total population of dwellings. The result is a housing market with little mobility, especially in the Greater Copenhagen area.

The skewed rent structure on the housing market has had an impact on the social housing sector (Skifter Andersen, 1999). Because of the difficulties getting access to other parts of the housing market, social housing has functioned as a last reserve satisfying persons with temporary and acute housing needs. After a period of time these persons move out again rendering the social composition of tenants unstable in some estates. In some social housing estates the relative proportion of socially disadvantaged persons has risen as a result of the municipal allocation scheme while persons with relatively more social and cultural resources have opted for other means of satisfying their housing needs like owner-occupied or co-operate housing.

5.3 ASSET MANAGEMENT PRACTICE

Danish social housing is organised as a publicly subsidised good delivered by private, non-profit housing associations and made subject to a substantial body of law regulating all areas of relevance to the key task of housing provision and maintenance. Thus, the strategic space of housing managers is limited, especially with respect to asset management as strategic market behaviour. However, if asset management is perceived as institutionalised procedures to optimise on the balances and trade-offs between housing quality, social objectives and cost-effectiveness, the Danish case may have more to offer. This section presents an overview of asset management practice in Danish social housing associations.[1]

5.3.1 Rent Setting

Rent-levels derive from building costs minus public subsidies. Older estates are cheaper compared to new ones because building costs have increased, and there is a general mismatch between price and quality in the social housing sector; some of the older and more attractive estates are also relatively cheap.

Social housing associations can propose rent increases but these have to be sanctioned by tenants. Rents are related to maintenance and renewal purposes, not to tenants' incomes. Since the housing sections are treated as financially independent units, costs associated with improvements and renewals are directly transferred to rents. Thus tenants are tempted to adopt a short-term perspective keeping rents down while housing administrators emphasise a long-term perspective. Normally, adequate appropriations for future improvements are accepted by all parties as a key element in the continuous modernisation of the housing stock. Social housing associations cannot raise rents.

Financial government is an important topic on the agenda of section boards. Budgetary powers have been decentralised to the level of tenants' assemblies, hereby strengthening the link between the democratic mode of organisation and the local choice of economic strategy. In many housing sections green accounting systems have been introduced describing the local and global environmental effects of activities in the estates (Jensen, 1999). Green accounting systems allow tenants to scrutinise past activities and life styles patterns in the search for potential reductions in levels of energy consumption and pollution and they potentially result in a noticeable reduction in running expenses while adding an input to the democratic processes in the estates.

[1] This description is based on 12 interviews with managing directors and tenants representatives from 5 social housing associations and the Federation of Non-profit Housing in Denmark as well as secondary sources.

5.3.2 Allocations

The internal and external waiting-list system is defined by central regulations and is administered by the housing associations. In co-operation with individual housing estates the associations decide on the structure of the system, i.e. the grouping of associations and estates in relation to specific lists.

Amongst tenants there is a high degree of satisfaction with the allocation system. Because rents depend upon costs at the time of construction typically the most attractive apartments are also the cheapest while the least attractive are the most expensive. Tenants tend to regard the waiting list system as a fair alternative to the market since everybody can sign up for a flat which they will eventually get it if they wait long enough, sometimes a lifetime. Nevertheless, for some years the allocation system is criticized, in particular for the little access of newcomers to the sector. Furthermore, the system further aggravates the segregation processes on the housing market; because of the division between internal and external waiting lists attractive apartments are only accessible to tenants who have signed up for them, often for a considerable number of years. Also, it is cost generating that some housing sections in effect function as transit sections from which tenants move on to housing sections of their priority according to internal waiting lists.

In many municipalities specific rules have been adopted with respect to entitlement criteria for access to social housing. In some municipalities regulations are in force with the explicit aim of excluding segments of the population (especially low-income households) from access to social housing (Skifter Andersen, 1999a). Hereby, the municipality avoids the burden of financing e.g. social transfers to social assistance receivers from another municipality. The local entitlement criteria allow for some degree of discrimination and inequality of access in order to impact the composition of the tenant population. The idea is that in the long-term perspective housing areas with a disproportionate high share of socially disadvantaged tenants will become socially 'stabilised' if the municipal housing allocation is directed towards other social housing areas. Tenancies are not related to the income levels of tenants, and there are no possibilities to evict tenants if their incomes increase relative to the level of the rent.

5.3.3 Sales

Up till now, the selling of dwellings has not been part of Danish associations' asset management and the proposal of the current government to allow for sale (see section 5.2.1) has met opposition from the sector. But, amongst housing professionals there is also a certain acceptance of a modified privatisation strategy. They oppose a right to buy, but accept a right to sell. The argument is put forward that sales should be orchestrated according to social criteria agreed upon by both social housing associations and local governments. It should not be the most attractive parts that are sold, hereby further residualising the sector, but dwellings in some of the worst off areas, increasing homeownership in these areas. Amongst tenants opinions differ, some are attracted to potential capital gains but most do not support a privatisation strategy.

Local governments share the scepticism to a large extent. They fear that the social allocation scheme is undermined if the number of available apartments drops. Out of the 470,000 housing units, 'social' allocations account for a total of 170,000 dwellings, and it is estimated that local governments require 14,000 vacancies annually to satisfy housing needs in the social allocation scheme (KL, 2002). According to the current legislation local government can veto a sale of social housing, and local governments advocate a model where the municipality in co-operation with the social housing associations decide whether to sell individual apartments or not.

From a management perspective a possible sale entails a number of uncertainties and challenges. The social housing stock will diminish, and though it constitutes a relatively large proportion of the housing market, it is likely to substantially reduce the portfolios of many associations. Will profits derived from sales be re-channelled into new social housing constructions? Will privatised housing-sections enter into new management agreements with existing social housing associations? How will a sale impact upon the administrative cultures of social housing associations? To what extent are the legislative barriers to a market-oriented behaviour of social housing associations loosened up, allowing the associations to compete with private landlords? No matter the likely outcomes to these questions, social housing associations discuss how to prepare for a future market situation in which they have to compete for their customers.

5.3.4 Maintenance and Renewal

Housing associations are required to set up long-term maintenance plans for all housing sections, and tenants and housing professionals carry out annual on-site evaluations of the estates and decide upon maintenance policies for each housing section. Thus, the maintenance strategies are developed as a mix of a centralised technical planning and budget system, in which the technical data on individual housing components are priced and systematised, and ongoing dialogues with tenants' representatives in all housing sections. Large-scale renewals are typically initiated by housing associations, but they have to be agreed upon by tenants. Funding for major renewal projects is derived from rent-increases and the National Building Fund when projects meet the funding criteria set forward by the fund. As mentioned above, the Urban Committee financed a nation-wide renewal process in the social housing sector in the 1990s.

Right of disposition ("Råderet")

In 1994 the concept of a 'disposition right' was introduced by law stipulating a set of rules within which tenants are allowed to dispose of their apartment, i.e. do maintenance and repairs etc. Before the introduction of this right, sometimes tenants made considerable alterations to their apartments and when they moved out housing associations demanded that the apartment be reversed to its original state. Typically this process was very costly and the lack of clear rules sparked off frequent conflicts.

With the disposition right, tenants are entitled to carry out some alterations without a specific permission, while other (major) physical changes to individual apartments have to be approved by the housing association. If acceptance is granted tenants are reimbursed. The regulation of the disposition right is somewhat bureaucratic, and tenants sometimes engage in do-it-yourself activities without consulting the administration.

5.3.5 Decision Making Process: Considerations, Methods and Strategies

Social landlords in Denmark are housing providers, not housing investors. Thus, they can not engage in portfolio investments, and they (still) cannot sell their housing stock. Furthermore, detailed regulations narrow down the scope for secondary activities such as external housing administration or technical service delivery. Potential financial losses are to be covered by tenants, and the legislative framework regulating the financial activities of housing associations is set up to prevent that tenants suffer financial penalties from mismanagement. Thus, most associations concentrate on primary activities pursuing a policy of economic stability and financial risk-avoidance.

Typically, housing upkeep is planned with a 10-year time horizon. This allows for accumulation of resources thereby preventing major fluctuations in rent levels. The housing association administers the financial flows and has a certain freedom to optimise on capital returns. In principle, the association can increase expenditure levels by employing more personnel, and by engaging in more activities without charging higher rents if it is successful in various profit-making activities. Profits are typically derived from higher return on capital or from fees derived from building activities. In a long-term perspective, financial returns primarily derive from increasing building values, and proper maintenance is thus a key factor.

In the late 1970s and early 1980s many housing associations experienced periods of continuous deficits, and most adopted a budget policy which enables the association to shift budgetary items from ad hoc posts to planned expenditures. Such policies increase overall savings and create a buffer zone against deficits and unforeseen costs. The orientation towards proactive planning and budgetary procedures was implemented when a legal obligation to do so was introduced around 1993. With the Social Housing Act of 1997 it was made mandatory to increase the financial transparency of the housing association and a number of separate accounts need to be made on specific issues such as the repair shop, the technical department etc.

Social investments

Because of the segregation process on the housing market relatively large proportions of residents in social housing estates are outside the labour market, and many residents stay in their housing area during daytime hours. The municipal housing scheme has further accelerated the segregation process resulting in a relative concentration of socially disadvantaged population groups in the least attractive housing sections with the largest number of voids. Typical social problems relate to alcoholism and drug-addiction, and a discrepancy exists between 'ordinary' tenants who see no reason to become socially involved "just because they live in social housing" and tenants with special difficulties and housing needs.

The Urban Committee Programme has to some extent slowed down the segregation process. The combination of rent reductions and a comprehensive process of physical renewal has had a positive impact on moving patterns, slowing down the vacation rate particularly amongst the well-off segments of the tenant population. Further, there is a tendency that more tenants have become involved as a result of the social development strategy, and the tenants' advisors scheme has been successful in reducing the overall levels of social problems especially in relation to youth-related issues (Skifter Andersen 1999b).

Methods for estate assessment

The Danish social housing stock is evaluated on a regular basis with the mandatory annual on-site evaluations of the estates. Also, housing administrations have access to data on the social composition of residents and they employ social indicators in their dealings with local government on the allocation issue. Some housing associations have developed rating systems that are presented and debated on the annual budget approval meetings with tenants. Such a rating system is for example composed of data on each estate with respect to rent level, savings, deficits, voids, upkeep, vacations and the general economic health of the housing estate.

The growing market-orientation is likely to result in increased visibility with respect to levels and composition of housing expenditures when housing administrations introduce procedures by which users get access to information about the price and quality of services provided. Each housing section pays an administrative fee to the housing association in return for services, the fee ranges from about DKK 1,800 to 2,500 (€ 242 to € 336) per housing unit per year. In principle, housing sections are free to choose administration but this rarely happens, typically local housing sections have cultural and political affiliations with their housing association.

According to law, social housing associations have to submit annual accounts to a database in the National Building Fund, a database that allows for comparison of costs between associations. With online access to these data housing associations can carry out benchmarking exercises and ideally promote transparency and rationalisations of administration costs through increased competition. Further, many associations consider differentiating between base-line products and individual services to accommodate differences in tenants' preferences allowing for a more individualised service profile. To the extent that such mechanisms of comparability are implemented the next step may be that tenants "shop" between different housing associations on the basis of comparison of costs, service levels etc.

Tenant involvement

Tenant involvement plays a key role in Danish social housing. At the level of individual housing estates, tenants are elected for section boards that deal with social, physical and financial aspects of the estates in co-operation with housing management. Further, tenants constitute the ruling majority of the housing council and the executive board of directors that appoint the housing director. It follows from the system of tenants' representation that tenants have a decisive say with respect to both day-to-day management decisions as well as long term strategic decision-making.

Many, especially the larger, housing associations employ social workers and tenants' advisors, and there is focus on the potential positive social return of the residents' participation in the administration and governance of housing estates. The combination of professional administrations and a decentralised participatory structure facilitates residents' mobilisation in the day-to-day physical and social development of the housing environment. Section boards function as conflict-mediators, and they facilitate tenants' participation in the production of collective goods in the estate as the cornerstones of the local participatory structure.

5.4 CONCLUSION

The social housing sector in Denmark consists of non-profit housing that is publicly subsidised in order to guarantee universal access to adequate housing. The public subsidy is accompanied by detailed public regulation that prohibits social housing associations from acting like a private enterprise or a private landlord. Despite variations in size and administrative cultures, all social housing associations are fairly similar in that they manage their assets according to detailed legislative guidelines and regulations. In this context, the scope for strategic, market-oriented asset management is limited. However, there is a longstanding tradition for comprehensive balancing and integration of social, functional and financial objectives in the sector. Also, with the new Liberal-Conservative government, a renewed political pressure towards privatisation and sale of social housing units is been exerted, forcing social housing associations to rethink their roles and strategies.

From a housing administrator's perspective it is a promise of continuity that tasks and purposes are clearly defined and housing activities comprehensively regulated by law. However, the degree of political control and the lack of administrative autonomy within the sector are experienced as drawbacks that makes it difficult for housing associations to take on the privatisation challenge and explore more market-oriented strategies. Also, for some decades demand for social housing has exceeded supply, and caretakers and administrative staff has tended to become less service-minded towards tenants. In this context, housing administrators and tenants' representatives continuously discuss how to modernise the sector. These discussions focus on especially three strategies:

A service-oriented strategy intended to increase the competitiveness of the sector in relation to the private housing market. A part of this strategy is to allow for individualised housing services and more flexibility with respect to the physical modernisation of apartments;

A democratisation strategy to support and strengthen the system of tenant's self-governance, the democratic dimension is a primary source of legitimacy and serves as a lay-input in the day-to-day management of housing estates;

A social cohesion strategy, which emphasizes the responsibility of the social sector to accommodate the housing needs of socially disadvantaged population groups, and the development of an institutional capacity to facilitate social cohesion in problem-ridden housing estates.

The above objectives tend to conflict with one another. From a market perspective, housing associations have an ambiguous relationship to the social obligation. Though there might be political and moral support of the social responsibility it is costly, and from a strategic point of view it is a future competitive advantage not to have estates that are too residualised. The democratic component is a cornerstone in the Danish model, and it is perceived as a value in itself. But from an administrative perspective the democratic processes are perceived as inefficient and the tenants as rather conservative; administrative leaders advocate liberalisation of rules regulating the sector while tenants' representatives only reluctantly change status quo. Nevertheless, the possible future sale of social housing is a wake-up call for housing associations. Furthermore, housing associations foresee that tenants will be more likely to shop between administrations in the future. To meet these challenges, housing associations push for reform initiatives that will allow for more flexibility in the organisation of housing services. Some of these initiatives include the following:

- A broadening of the scope for tenants' individual modernisation of flats, and tenants should be able to capitalise on improvements when they move;
- A differentiation of administrative services that allows tenants to choose between different service levels;
- Social housing associations should be allowed to embark upon administration of dwellings and institutions outside of the social housing sector on market terms;
- If social housing is to be sold this process should be governed by the sector itself in co-operation with local authorities, and capital gains rechannelled into new social housing constructions.

With the increased degree of self-financing in the sector the rationale for comprehensive public regulation weakens, and maybe the sector as we know it today will undergo radical changes in the foreseeable future.

REFERENCES

Boligministeriet (1988a) Cooperative housing in Demark.

Boligministeriet (1988b) Non-profit housing in Denmark.

Boligministeriet (1996) Rapport fra arbejdsgruppen vedr. 1.gangsplacering af flygtninge. Flygtninge og indvandreres bosætningsmønstre. Boligministeriet.

Bolig og Byministeriet (1998) Oversigt over ansøgninger om forsøg med udlejning og anvisning af almene boliger, 8. Kontor, notat 9.3.1998.

Bolig og Byministeriet (1999) Bygge- og boligpolitisk oversigt 1998–1999.

Bolig og Byministeriet (2001) Finansministeriet. Økonomiministeriet. Budgetanalyse af den almene sektor.

Engberg, L.A., Gottschalk G. and Pedersen, D.O. (1999) Beboersammensætning og tilfredshed i fem boligområder i Odense; en sammenlignende undersøgelse af forholdene i Blangstedgård, Seden Syd samt tre områder i Vollsmose, SBI rapport 325.

Groes, N. (ed.) (2003) Salg af almene boliger—Vejen frem? En antologi om dansk boligpolitik, AKF Forlaget 2003.

Hansen, K.E. and Skifter Andersen, H. (1999) Strategies for public regulation of urban renewal and housing rehabilitation in Denmark, in: Skifter Andersen and Leather (eds.), Housing Renewal in Europe, The Policy Press.

Hummelgaard et al. (1997) Udsatte boligområder i Danmark, AKF-rapport, København, AKF Forlaget.

Jensen, L. (1997) Demokratiforestillinger i den almennyttige sektor, Ph.D.-afhandling, Institut for Statskundskab, Københavns Universitet.

Jensen, O.M. (1999) Grønt regnskab for boligområder, SBI-rapport 303.

Jørgensen, H. (1991) The Administration of labour market policy in Denmark in the 1980s, in: Knudsen, T. (ed.), Welfare administration in Denmark, Copenhagen, Ministry of Finance.

KL (2002) En hel løsning? Et debatoplæg om salg af almene boliger, KL: Kommuneinformation.

Landsbyggefonden (2002) LBF-orienterer nr. 313.

Landsbyggefonden (2003) Notat til Landsbyggefonden orienterer nr. 317 af 28.

Lind, O. and Møller J. (1994) FolkeBolig. BoligFolk. Politik og praksis i boligbevægelsens historie, Boligselskabernes Landsforening.

Lovforslag nr. L 46: Forslag til Lov om ændring af lov om almene boliger samt støttede private andelsboliger m.v., By- og Boligministeriet, 8. Kontor, j.nr. B-5211-18.

Norvig Larsen, J. (1999) Citizen involvement in community revitalisation projects, working paper, SBI.

Odense kommune (1998) Tværmagistral Arbejdsgruppe vedr. ny bosætningspolitik i Odense Kommune.

Pedersen, D.O. (1998) Improving local social housing competitiveness—efforts and effects. Paper to the Conference on Housing Futures: Renewal, Sustainability and Innovation, ENHR, Cardiff, University of Wales, UK 7–11 September 1998, The Danish Building Research Institute.

Skifter Andersen, H. and Ærø, T. (1997) Det boligsociale danmarkskort. Indikatorer på segregation og boligsociale problemer i kommunerne, SBI-rapport 287.

Skifter Andersen, H. (1999a) Boligpolitik i Danmark, working paper, SBI.

Skifter Andersen, H. (1999b) Byudvalgets indsats 1993–98. Sammenfattende evaluering, SBI.

Vestergaard, H. et al. (1997) De 8 modelområder. Evaluering af et Byudvalgsinitiativ, SBI-rapport 288.

Vestergaard, H. (1998a) Troubled housing estates in Denmark, in: Madanipour, Cars & Allen (ed.), Social exclusion in European cities; processes, experiences and responses, Regional Studies Association, Regional Policy and Development Series 23, pp. 115–130.

Vestergaard, H. (1998b) Social exclusion in European neighbourhoods—processes, experiences and responses; The National Research Report—Denmark, working paper, SBI.

6 ENGLAND

ANDREW THOMAS and VINCENT GRUIS

6.1 SOCIAL LANDLORDS IN ENGLAND

In Britain the term 'social housing' or 'social rent housing' is used to describe two distinct but overlapping sectors, local authorities and housing associations. Local authorities can be regarded as public housing companies. Housing associations are voluntary not-for-profit organisations, but as Registered Social Landlords (RSLs) have public objectives and operate under close government regulation.

Local authorities were the main providers of social rented housing in England throughout the 20th century, and despite programmes of sale and transfer, they continue to own and manage 60% of the social rented stock. It is, however, a much reduced sector. In March 2003 there were 2.457 million local authority dwellings, compared with a peak figure of 5.140 million in 1979. While local housing authorities vary enormously, the average stock in ownership is much larger than RSLs. It is essentially a purpose built and aging stock, with about three-quarters now more than 40 years old. Despite differential patterns of sale, the proportion of flats still represents only 40% of the stock, and the majority of this is low rise (ODPM, 2003c). In comparison, the RSL stock is newer, and nearly half the stock is flats, again largely low rise, but with more conversions. There are many more RSLs than local authorities, and while the largest local housing authority is larger than any RSL, the RSLs vary enormously in size and characteristics. Nearly 80% of housing associations own just 8% of the stock, with 1,513 associations owning less than 1,000 homes. At the other end of the scale, the ten largest associations own 13% of the total stock, with an average ownership of 23,700 units.

Diversity has been the key characteristic of the RSL sector. There are "a wide range of quite different organisations, varying from ancient almshouses trusts and Victorian charitable foundations to self-build co-ops and former local authority housing departments" (Malpass, 2001, p. 227). But this latter category has led to significant changes in the sector. Over the past ten years the national policy of stock transfer from local authorities has altered the composition of housing associations: increasing the number of larger associations; increasing the number with a concentration of stock; changing the balance of stock characteristics in favour of a purpose built stock, with more estates

Vincent Gruis and Nico Nieboer (eds.),
Asset Management in the Social Rented Sector, 97–118.

Table 6.1 Division of the housing stock by tenure in England, 31 March 2003

	Social rented	Private rented	Owner-occupied
Percentage of total housing stock	19%	10%	71%

Source: ODPM (www.odpm.gov.uk).

Table 6.2 Some characteristics of social landlords in England, 31 March 2003

Characteristic	Local authorities	RSLs
Number of dwellings	2,457,000	1,621,000
Number of landlords	241	1,925
Dwellings per landlord (average)	10,195	842
% of total housing stock	11%	8%
% of social rented stock	60%	40%

Source: ODPM (www.odpm.gov.uk).

and less dispersed stock; and increasing the proportion of staff coming from a local authority rather than 'voluntary sector' background.

When set within an international context, social landlords in England have a relatively strong focus on housing for low-income households (see Van der Heijden, 2002). There is, as yet, very little market rent housing provided by housing associations (or strictly speaking, by their subsidiaries), but the future opportunity for private developers to compete for Housing Corporation funding may provide additional impetus. The provision for this is part of the current Housing Bill, and seems to be driven by a view that the involvement of private developers will bring fresh ideas, creativity and flexibility to the delivery of affordable housing. As things stand, private developers will be eligible for government grants but will not be subject to the same regulatory regime as RSLs. One possible result is reverse competition from housing association development teams to provide a mix of building for conventional sale, affordable housing for sale, shared ownership and conventional social rented housing. There is evidence of interest from some associations to build-for-sale as part of a mechanism to cross-subsidise social housing activities, and both home ownership and market rent subsidiaries reflect this strategy (see e.g. Barker, 2003).

6.2 NATIONAL CONTEXT FOR ASSET MANAGEMENT

Local authorities and housing associations operate within different statutory, regulatory and funding frameworks. They started independently and developed on different legislative paths. Nevertheless, both parts of the social rented sector share key similarities. The stock has been provided through public subsidy and is subject to regulation. Rents are set below market rent levels, and allocation was, and broadly still is, linked to the concept of need. The key differences are that local authority stock is geographically more concentrated, is purpose built, and has management accountability through the local political process.

Both local authorities and housing associations are under pressure to develop business-like approaches towards housing management. From 2001/02, local authorities in England began to operate under a new financial framework. Resource accounting was introduced to reflect capital employed and asset depreciation. The associated introduction of business plans (DETR, 2000b) is part of the process of encouraging authorities to make better use of their housing assets. These business plans draw on an investment plan informed by stock condition surveys carried out in accordance with good practice guidance, and in the future may result in more sophisticated approaches to asset management modelling.

The other main driver of business plans is the policy that local authorities should separate their landlord function from their strategic role. The message is that investment plans need to be funded, and that there are just three options: stock transfer (LSVT); arms length management organisations (ALMOs); and private finance initiatives (PFI). If the option appraisal results in a LSVT, then the investment plan will be the subject of more detailed work to inform the valuation, offer document and the development of a procurement strategy. Whether for transfer, ALMO or stock retention, the investment plan will form the basis of future business planning.

Housing associations operate in a parallel world. Though not affected by the same regulatory regime as local authorities, aspects of business planning have been part of their operational framework for a number of years. They are subject to the Housing Corporation's Regulatory Code, which requires them to operate viable businesses with adequate resources to meet current and future business and financial commitments. The business plan must demonstrate priorities and strategies for achieving the RSL's objectives, and the plan must be informed by asset management information which is reviewed regularly (Housing Corporation, 2002). All business plans are assessed by a 'lead regulator' to check compliance with regulatory requirements.

The recent government agenda for social housing policy was set by the Housing Green Paper (DETR, 2000a). This covered the promotion of owner-occupation, the delivery of affordable housing, changes to the allocation of social housing and the convergence of local authority and housing association rents. In terms of asset management there are three points to make about the Green Paper. First, it emphasised the strategic role of local authorities and the need to prepare housing strategy statements which are cross-tenure, evidence based, and consistent with national, regional and sub-regional objectives. From July 2003 Regional Housing Boards acquired responsibility for preparing regional housing strategies. From April 2004 the system of allocating funds for housing investment will change to reflect these regional strategies. In each region there will be a 'single pot' for housing investment, replacing the existing local authority housing investment programme (HIP) and Housing Corporation Approved Development Programme (ADP). The Regional Housing Boards will advise on the distribution of spending, though for now actual allocations will flow through existing channels.

Second, the Green Paper sought to improve the quality of social housing. Past levels of investment in stock maintenance were acknowledged to have been inadequate, but the promise of additional money came together with a new Housing Inspectorate to assess authorities' housing investment programmes, and a renewed emphasis on stock transfer from local authorities to (usually new) housing associations.

Third, the Green Paper introduced what was to be known as the Decent Homes Standard. This was little more than a line in the Green Paper. Reflecting the Government's taste for both jargon and targets there was to be 'a step change in the quality of the stock, with all social housing reaching a decent standard within ten years' (DETR, 2000a, p. 17). This was adopted as a target in the Government's July 2000 Spending Review, and in the 2002 Review was extended to cover private sector housing occupied by vulnerable people (HM Treasury, 2003, p. 14). To deliver decent homes, social landlords are required to quantify the level of non-decent housing in their stock, develop an investment strategy to tackle this, and measure progress towards the elimination of non-decency by 2010. The business plans of local authorities are assessed by Government Office to ensure these plans are capable of delivering the decent homes agenda. Housing associations have a similar requirement to demonstrate this to the Housing Corporation (DTLR, 2002).

The government action plan which followed the Green Paper appeared, with little sense of urgency, in 2003 under the title 'Sustainable Communities: Building for the future' (ODPM, 2003). This document, often referred to as the Community Plan, emphasised that achieving decent homes was at the heart of sustainable communities, with 1.5 million households estimated to live in social housing that failed the standard. (ODPM, 2003, p. 13). Repeating the target that all social housing should be decent by 2010, the Community Plan made it clear that all local housing authorities would be required to produce option appraisals demonstrating their ability to make the necessary investment in their stock, both to achieve decent homes, and to meet the aspirations of tenants within a fundable business plan and that the option appraisals would have to be signed-off by Government Office (the regional arm of ODPM) by July 2005.[1]

6.2.1 Regulations

General administrative regulations

The general administrative regulations have been laid down in various Housing Acts. The consolidating legislation of 1985 took the form of a Housing Act, Housing Association Act and Landlord and Tenant Act. Since then the Housing Act 1988 introduced assured tenancies, tenants' choice, large scale voluntary transfer and Housing Action Trusts. The Local Government and Housing Act 1989 followed with almost unseemly haste, introducing a new system of local authority subsidy. The titles became even longer with the Leasehold Reform, Housing and Urban Development Act 1993, which introduced the Right to Repair and the Right to Compensation for improvements carried out by tenants. The Housing Act 1996 turned its attention to homelessness and anti-social behaviour. After that, rather strangely, the date of the incoming Labour government is impossible to identify from housing legislation. They came to power in 1997 but did not produce a policy paper until 2000. Instead, the housing agenda was dominated by managerialism—issues like Best Value (Local Government Act 1999), and business planning (reforms to accounting practice and subsidy calculation requiring administrative reform or detailed legislative changes). The implications of these acts and legislations for the various asset management activities are discussed below.

Rent setting

Local authorities have traditionally been given autonomy in setting the rents of individual dwellings in order to raise the total rent income required to meet their expenditure on public housing. Prior to 1989 housing association rents were determined independently by the rent officer service. Following the 1988 Housing Act, housing associations had the freedom to determine their rents on new assured tenancies.

The past legislative context has lead to substantial (and some irrational) differences in rents between sectors and between individual landlords within sectors. Housing association rents in general are higher, at an average of £60 a week in March 2003 compared with just under £50 for English local

[1] The requirement to have signed-off option appraisals is an example of the way English government talks about devolution but behaves in a very centralist fashion, micro-managing the role of local authorities.

authorities. The government is now operating a policy aimed at achieving similar rents for similar properties across the social rented sector. Target rents are related to a formula where the weekly rent is equal to 70% of the average rent for the local authority or RSL sector multiplied by relative county earnings and weighted by number of bedrooms, plus 30% of the average rent for the sector multiplied by relative property value (DETR, 2000a). The effects of this rent convergence depend on local property values and existing rent levels, but some associations will experience a real reduction in rents over the ten year implementation period, while overall, local authority rents are expected to increase.

Allocations

Social housing in Britain has conventionally been allocated on the basis of need. This has usually been assessed on a points system, with the house allocated to the applicant with the highest points score. There is no legal definition of housing need, but there is guidance for local authorities (ODPM, 2002). This guidance refers to the way local authorities must allocate housing under Part 6 of the Housing Act 1996 (as amended by the Homelessness Act 2002). It is relevant to RSLs in that under the Housing Act 1996, RSLs have a duty to cooperate with housing authorities 'to such extent as is reasonable'. But RSLs are also subject to the Housing Corporation's Regulatory Code. According to recent regulatory guidance from the Housing Corporation (2002), allocations:

Must be – responsive to local authority policies;
Must be – non-discriminatory;
Must be – responsive to demand;
Have to – ensure sustainable communities;
Should – co-operate with local authority homelessness functions;
Should – where reasonable, provide for local authority nominations;
Should – take account of transfer applications, including those from other associations.

Despite the legislation and guidance, in general terms social landlords in England have a substantial degree of autonomy in deciding who they house. A local authority allocation policy must consider all applications, without blanket exclusions, but in determining relative priorities they are explicitly permitted to favour local connection, to exclude on the basis of anti-social behaviour, and to give lower priority to an applicant who is financially able to secure alternative accommodation through market renting or owner-occupation (ODPM, 2002).

Under the Housing Acts, local authority properties are let on the basis of secured tenancies. The tenants' rights and responsibilities are set out in the

Council Tenants Charter. The most important points are that they have a permanent tenancy, with rights of succession and, with limitations and exceptions, to discount purchase. The Council must maintain the property internally and externally, with tenants responsible only for wear and tear, decoration and furnishing. The tenant must conform to the tenancy agreement, pay the rent and not behave in an anti-social manner.

Housing associations let property on assured tenancies. Only those living in their homes before 1989 will be secure tenants. In practice this is not usually important. Assured tenancies were introduced in the Housing Act 1988 and were framed for private sector landlords. Compared with secure tenants, assured tenants under the Act have fewer rights of succession, assignment and consultation. However the Act provided that the Housing Corporation be given power of guidance known as the Tenants' Guarantee. Under this, housing associations will normally operate a model tenancy agreement which is equivalent to the statutory rights of secure tenants.

Sales

The sale of local authority stock to tenants began with the introduction of the 'Right to Buy' in the 1980 Housing Act, permitting tenants to purchase their dwelling at a discount, though if they sold within three years, some or all of the discount was repayable. This legislation excluded charitable RSLs, but included secure tenants of other housing associations and tenants transferring from local authorities.

Significant transfer of dwellings from local authorities to housing associations has also occurring through 'large scale voluntary transfers' (LSVT's). Through these transfers, housing is sold by the local authority to an existing or (usually) new housing association. These privately funded sales require a detailed assessment of the quality, market position and financial prospects of the housing stock, with the business plan and underpinning investment strategy being significant factors in the development of asset management practice.

Maintenance and renewal

While local authorities and RSLs both have a responsibility to maintain and renew their stock and provide a repair service, they operate under very different frameworks. Both types of landlords are subject to the objectives of the decent home standard, but RSLs must fund all their maintenance and investment plans from their own resources. They can raise private finance in order to fund development programmes or to reinvest in their existing stock. As part of the Housing Corporation's Regulatory Code, associations are required to monitor their progress towards achieving the decent homes standard. Furthermore, RSLs must ensure that:

– their homes are well maintained and in a lettable condition;
– maintenance is carried out effectively and responsively and in ways that reflect residents' preferences;
– necessary investment in the future of the stock is made a key priority.

In contrast to housing associations, local authority investment continues to be eligible for central government support. Local authority Housing Investment Programme (HIP) allocations are based on government assessment of requirements for housing investment, informed by local housing strategies, and from 2004, by the priorities set out by Regional Housing Boards. There are two elements. One is a matching grant for disabled facilities, but the more important is the Annual Capital Guideline (ACG), which involves:

– Receipts Taken Into Account (RTIA)—the capital expenditure central government assumes each local authority can finance from their usable capital receipts;
– Basic Credit Approval (BCA)—before the start of the financial year the government issues each authority with a single BCA to be used to support any service including housing, and represents the local authority's borrowing limit;
– Supplementary Credit Approvals (SCA)—for capital expenditure on specific projects.

A Housing Revenue Account (HRA) is operated by every local housing authority, consisting of:

– income from rents and housing subsidy—which has elements for rent rebate, housing management and stock maintenance;
– expenditure on the management and maintenance of the council housing stock.

In April 2001 the government introduced a Major Repair Allowance (MRA) as part of the HRA. The MRA is the subsidy allowance paid by government to represent the cost of maintaining the current condition of the stock. It was introduced as part of a new financial framework for local authority housing and is seen as a means of accounting for the depreciation of the stock. In this sense it is part of the process of converting the HRA to reflect both the cost of capital employed in housing assets and the cost associated with the deterioration of these assets.

6.2.2 Supervision

The ODPM is the government department responsible for housing policy. Regulation of the local authority sector is partly through Government Office and partly through the Audit Commission[2]. Housing associations are also inspected by the Audit Commission, but with the Housing Corporation acting as the lead regulator.

The Housing Corporation is a government sponsored body originally established under the Housing Act 1964 to promote housing associations and co-operatives. It regulates RSLs in England, and since 1974 has been the major funder of RSL capital programmes through the Approved Development Programme (ADP).

Recently, in 2003, the Housing Corporation introduced Housing Corporation Assessments (HCAs). These are a public statement of the Corporation's view of every RSL under three headings: viability, governance and management. Each year all associations with more than 250 dwellings are expected to provide a five year financial projection, and these are used to inform the Corporation's annual viability review (AVR). AVR's are based on credit-rating methodology and are treated as commercially confidential, but are reflected in the viability section of the HCA. Longer financial projections are required of RSLs planning to commit to major development programmes.

In 2001 the Housing Corporation emerged from its annual five year review with confirmation that its focus on regulation and investment was to remain, but that there should have separate directors for the two functions. Reflecting this, the Corporation published a Corporate Strategy (Housing Corporation, 2001). However, the inspection component of its regulatory role was then lost when the Audit Commission and Housing Corporation's inspection services were merged, under the Audit Commission, to form the new Housing Inspectorate. For RSLs the Housing Inspectorate reports to the Housing Corporate on any matters considered to be of regulatory concern (Audit Commission, 2003).

6.2.3 Financial Support

For the best part of the whole of the last century it was local authorities who were the agents for delivering government housing policy. The present system of financial support was introduced in the Local Government and Housing Act 1989. This created a ring-fenced Housing Revenue Account (HRA), a new subsidy system and new controls on borrowing which have, with modification, remained in place until the Local Government Act 2003. We have seen that capital funding under the 1989 Act was through the Housing Investment Programme (HIP) and more recently the Major Repairs Allowance (MRA),

[2] The Audit Commission is another non-government agency: an independent body responsible for reviewing the spending of public money in local government, housing, health and criminal justice services. Its inspection reports in the local authority sector are in the context of CPA, with the resolution of service delivery failure being with ODPM.

and the main elements of this survived the 2003 Act. Local authorities can renovate their own stock, support the renovation and development programmes of RSLs, and provide renovation grants to the private sector.

The introduction of the 1988 Housing Act meant a significant change in financial support for housing associations. Public finance was altered through a system of mixed finance, with operational costs and risks transferred to individual associations. On the revenue side grants were effectively discontinued, requiring associations to balance their own books. RSLs were therefore more exposed to development risk. The level of grant was fixed before the scheme went on site, the balance funded, usually by borrowing, and future maintenance costs had to be met from income. However, despite this exposure to the private market, associations have not become financially independent, since substantial public development subsidy remains. The Housing Corporation still funds investment in new building and renovation through an Approved Development Programme (ADP), though this is destined to be replaced by single regional pots under the control of Regional Boards.

Housing associations borrow private finance on commercial criteria. The loans are secured against assets, and the interest terms defined in part by the credit rating of the borrower. There is no government guarantee beyond the fact that the lending institutions know the social rent sector is highly regulated, and that a failing RSL would have to be rescued in the interests of the tenants.

As well as production subsidies, consumption subsidies exist for social rented tenants in the form of Housing Benefit. This is an means-tested income subsidy to help low income groups pay their rent. The calculation is complex, but all of the eligible rent is paid for households whose assessed net income is less than a threshold calculated in relation to household type, age and circumstances.

There can be considerable differences between the financial position of local authorities and housing associations, caused by factors such as past patterns of central government funding, levels of debt, the local housing market, the condition of their stock, and the financial threats posed by rent restructuring. The strategies to meet these financial challenges include LSVTs, ALMOs and mergers between housing associations.

6.2.4 Housing Market

In the context of asset management in the social rented sector, we focus here on two aspects of the English housing market: first, the need for affordable housing and second, low or changing demand.

As noted by Keenan et al. (1999) there is a tension between the facts of low demand and the orthodoxy of national housing shortages in the context of a projected 4.4 million new households by 2016. The basis of household projections is of course disputed, but government policy takes an assumed annual growth rate of 155,000 households. This underpins arguments in the Communities Plan (ODPM, 2003) for growth areas in the south of England to increase the supply of housing, and specifically to address the need for affordable housing (the particular focus of the 2003 Barker Review of Housing Supply).

The supply of affordable housing is not just an issue for the urban south and rural areas, but for this discussion it has to be overlaid with regional patterns of demand. The Communities Plan recognised issues of low demand and abandonment, and subsequently published a separate plan (ODPM, 2004). The Government accepts estimates that there are around one million homes affected by low demand and abandonment, spread over 120 local authorities in the English North and the Midlands, including both social and private sector housing (ODPM, 2003). The speed of this market decline has been rapid, or at least went undetected until the second half of the 1990s. The explanations put forward for low or changing demand have been varied, and quite properly have emphasised the complexity and variability of local housing markets. Regional economic decline has been a factor, together with migration patterns within sub-regional housing markets (see Murie, 2002 for a fuller discussion). Outward migration affecting some of the older conurbations has left a stock of less popular housing which is close to being socially if not physically obsolete, a trend which was not managed by regional and urban regeneration policies over some twenty years from the mid 1970s.

Thus, although in general the demand for social rented housing exceeds the supply, social landlords in some areas are faced by rising rent arrears, higher level of voids and increasing turnover-rates. Describing this, Brown et al. (2002, p. 4) point out that this is "having a negative effect on the financial and management performance of landlords ... Yet, at the same time, there are issues of lack of affordable accommodation in some areas, where there are often high land values, making it difficult for housing associations to develop. Although this is often portrayed as a regional division between high and low demand, the reality is more complex, with 'hot spots' and long-term voids in adjacent neighbourhoods (see Blackaby, 2000)".

Social landlords are therefore faced with low demand for relatively new stock, while in an adjoining area there could be demand for affordable housing which is difficult to meet. These circumstances set demanding challenges for asset management. Case studies by Larkin (2000, p. 13) have shown that while the majority of stock "... is in good condition and well located ... it is also clear from the case studies that a substantial proportion of the stock of housing associations presents asset management issues which require a more fundamental appraisal of the options available". Much the same point could be made about the stock of many local authorities. Particularly in the older conurbations social landlords are operating in an environment of managed decline or long term progressive renewal, with the need to market their stock in competition with low cost home ownership and other rented options both inside and outside the area. The challenge for asset management will be the development of appraisal techniques which take into account both the views and aspirations of local people, and at the same time can cope with the uncertainty of local market behaviour in the context of attempts to achieve market renewal across sub-regional housing markets.

6.3 ASSET MANAGEMENT PRACTICE

There is widespread interest in asset management in the English social rented sector. In response to a more market-orientated approach to the management of social housing, questions are raised about which units can be sold, where homes might be built, which stock needs to be renovated or offered to a different target market. To support these strategic housing policies, performance analysis and option appraisal are activities which are becoming more common. There is now general recognition in the social rented sector that an asset management strategy is a requirement in order to inform business decisions. It is an essential tool to help the business operate efficiently and meet the needs of current and future residents. To quote a recent NHF report:

> *'Without properly costed asset management plans the business plan of a registered social landlord is a shot in the dark. The board, the chief executive and the senior management team should be devoting enough time to understanding their existing asset base and making decisions on investment and sales. Asset management planning is fundamental to the future viability of social housing in the UK. In some cases, decisions on asset management will be critical for the future viability of the organisation itself.' (NHF, 2001:1)*

In this section we discuss whether this is reflected in current the practice of English social landlords.

6.3.1 Rent Setting

For much of their history, local authorities in the UK have operated rent pooling policies under which the primary objective was to raise sufficient rent income to meet their expenditure on public housing. As a result, the rents of local authority dwellings have not necessarily reflected their current costs, historic costs or current value. Both pooling and the rent setting methods that local authorities have employed have tended to result in rent structures with limited differentials between different types of property and between different locations within local authority areas. Before the recent rent-restructuring, a survey by Walker and Marsh (2000) shows that point systems were the most frequently used method (by 54% of the local authorities in 1998/99). Within this system, points were awarded for dwelling characteristics such as type, age, size and location.

Under the 1988 Housing Act housing associations were able to determine their own rents in the light of their financial commitments and plans for future housing and service development. Given that housing associations and local authorities have had discretion to set the rent of individual properties it is not surprising that similar properties in the same areas might have quite different rents (Walker and Marsh, 2003).

Within the new regulatory framework, we have seen that rents will be largely determined by the nationally applied rent setting formula, which establishes target rents on the basis of local property values and local earnings. The specific concern for asset management is that rent restructuring makes it even more difficult to influence the demand for a particular type of property or area. While being asked to operate in an essentially market framework of business planning and funding, when it comes to the pricing of its product, social landlords in England have little freedom to increase the rent of stock where demand is high, or to lower the rent in areas where demand is low.

6.3.2 Allocations

There is considerable variation in the detail of allocation systems among social landlords in England. Most, however, have policies designed to give long term help to those in the greatest housing need. Traditional allocations methods have centred on points-based systems and housing registers. Need is defined by decision-makers (e.g. housing professionals/officers, councillors, and board members) in widely different ways. Need is expressed in terms of the accumulation of points which are awarded on the basis of current housing situation (e.g. overcrowding, affordability, occurrence of harassment) and specific factors (such as medical priorities). The relative number of points given for a particular factor can vary considerably. The development of these policies will depend on the interplay between many factors including the nature of the local housing market and local political considerations. Areas with relatively high demand for social housing are likely to operate more restrictive and less open access policies that incorporate, for example, income eligibility criteria and 'local connections' (including residential requirements). Substantial differences can also exist between adjoining local authorities even though they may cover the same housing market. This is illustrated in Table 6.3 which shows how the principles for the award of points for specific factors vary between three local authorities within the same sub-region (Brown, 2000). Following the Dutch experience, choice-based letting mechanisms have also been introduced in England, but these are not yet widely applied (see also section 10.3.2).

Table 6.3 Local authority allocation systems—principles for awarding of points

Factor	Local Authority A	Local Authority B	Local Authority C
Personal Situation: Harassment	Defines harassment as actual domestic and racial abuse	Defines harassment as actual and threatened harassment	Defines harassment as actual, threatened or perceived abuse
Social Needs: Medical Issues	Assessed by independent medical officers	Medical officers advice taken into account by housing allocation staff	Defines medical needs in a different category to personal and social needs—use of medical advisory office
Housing Situation: Overcrowding	Points awarded for insufficient bedrooms with doubling of points for each lack of bedroom	A set number of points awarded for insufficient bedrooms	Range of points awarded depending on circumstances. Officers have discretion to award additional points
Efficient Use of Council Stock	Points awarded if household is currently under occupying a council house	Points awarded if household is under occupying certain types of property (e.g. adapted houses)	No points awarded, but issue may be considered in the allocations process
Financial Circumstances	No use of income criteria	No use of income criteria, but income limit affects access to housing register	Points awarded of assessment by officers indicates an inability to afford existing home

Source: Brown (2000).

Recent research on RSL lettings policies (Pawson and Mullins, 2003) found that, excluding LSVT RSLs, a third of the remaining large associations did not have their own waiting lists but simply processed transfer requests and prioritised local authority nominations and other referrals. Many of the other RSLs were involved in locally based partnerships such as common housing registers and choice based letting schemes. The general conclusion was that the lettings function of an RSL offered very little independence in the context of both their duty to co-operate with local authorities and the increased importance of common registers and choice based schemes. The exception to this is in areas of low demand, where RSLs do have some freedom to alter their criteria within the framework of local lettings plans.

6.3.3 Sales

As yet, the selective sale of property has not been a key feature of local authority or housing association asset management strategies. Larkin (2000:37) suggests that the rate of sale in the RSL sector is relatively low given the size of the sector and the range of 'asset challenges' it faces. In future, however, there is likely to be growing interest in disposal. Newley (2002) amongst others has noted that the restrictions on rent setting plus the demand to improve quality under Decent Homes might stimulate RSLs to develop a more proactive approach to stock disposal. However, there will still be reluctance to dispose of stock on any scale. The larger RSLs have a culture of growth, and selling-off the family silver is not an immediately attractive option even where the assets are proved to be nickel-silver at best. There are also the attitudes of the local authority and Housing Corporation to be taken into account, together with regulatory codes.

Sales of local authority stock are largely determined by the legal context (the Right to Buy) and the political pressure underpinning stock transfer. Since 1980, more than half a million homes have been sold as part of the Right to Buy. LSVTs have also been substantial. By February 2003, 143 local authorities had carried out 180 transfers of a total of 738,000 homes, representing 18% of the 4.2 million homes owned by local authorities at the start of the transfer programme in 1988 (NAO, 2003).

6.3.4 Maintenance and Renewal

We saw in section 6.2.1 that the maintenance of local authority stock is influenced significantly by central government guidelines and the major repairs allowance paid as part of the new financial framework. This said, the relatively low standard set by Decent Homes is not always fundable, with option appraisal solutions leading to stock transfer or ALMOs as the mechanism for delivering additional investment.

For housing associations, the 1988 Housing Act exposed them to private sector finance under a mixed funding regime, with no government grants to cover future maintenance costs. The importance of asset management has therefore been given prominence because associations must make adequate financial provision from their own resources for the future repair and maintenance of their stock (NHF, 1997). Risk management has become even more pressing in the context of 'rent convergence', which is a serious threat to the financial resources of some associations. In this climate, if associations are to manage their financial affairs prudently, they need to understand the future shape of stock investment requirements. In turn this requires information on which to plan, and stock condition surveys have been undertaken by many associations over the last ten years, initially under guidance from the NFHA (1994), and more recently DETR (2000a). Larkin (2000) indicates that methodologies for stock condition assessment are fairly well established in theory, but that the state of play within housing associations varies. Furthermore, all of the case study associations that Larkin examined recognised the need for taking on a more systematic approach to planned maintenance, mainly to integrate planned maintenance with a comprehensive approach towards asset management, ensuring that the budget is spent on planned rather than responsive and void maintenance and integrating priorities for planned maintenance with estate assessment, tenant involvement and marketing approaches rather than relying solely on stock condition assessments.

6.3.5 Decision-Making Process: Considerations, Methods and Strategies

In our discussion of the decision-making process we will focus mostly on housing associations for two reasons. First, although in general discussion about English landlords is hampered by the lack of systematic information, there are a number of case studies available from the housing association sector (Larkin, 2000; William Sutton Trust, 2000 and Newey, 2002). Second, because of their early exposure to mixed-funding regimes and the impetus of LSVT asset management, it may be expected that innovations are most likely to occur in this sector.[3] This would reflect the role housing associations acquired in the 1980s as the government's preferred provider of social housing, among other reasons because they were seen as more detached from government and were expected to operate more like 'real' players in the market (Walker and Van der Zon, 2000).

Although English associations are under pressure to operate in a more business-like manner, there is little evidence that they have (already) adopted a market-orientated approach within in the specific area of their asset management. According to Larkin (2000, p. 5) "in general, the assessment is that associations are now only coming to focus on the performance, location and viability of their existing stock". The case studies by Larkin (2000) also show little evidence of strategic differentiation of the portfolio or variation of maintenance according to the market position of dwellings and household preferences.

[3] A preliminary assessment by Walker (2001) suggests that the impact of New Public Management and the adoption of it's associated management approaches is stronger within housing associations than within local authorities.

In terms of the prerequisites, the majority of social landlords are moving towards regular stock condition surveys. This is a requirement of local authority option appraisal, and something the Housing Corporation look at quite carefully in their regulatory role. However, the maintenance and analysis of this data is still at an early stage of development. Large social landlords will all have integrated housing management systems, and these will usually have some kind of planned maintenance module. These modules allow programmes of work to be assembled against one or more properties, costed against a schedule of work and linked to a contractor, with ordering and invoicing interfaces to a general ledger. The limitations are that the primary purpose of these planned maintenance modules is to create and maintain data. They are largely unaware of business rules and are independent of business processes. Typically there is no functionality within this type of module for forecasting future costs or for undertaking 'what if' appraisal of spending plans. Recognising this limitation, a number of PC products have been developed with specialist asset management aspirations. These are improving in sophistication, and as they do the gap is widening between their functionality and the capacity of clients to exploit the potential.

We have seen that business plans and the requirements of option appraisal mean that high level strategic or business planning activity is now relatively well established in England. Below this, there has been a limited amount of work on portfolio classification: "Some associations have developed systematic approaches to categorising their stock for asset management purposes, and then attaching strategies to particular categories" (see also the case of the William Sutton Trust, 2000). "Others [most] have adopted more pragmatic approaches, usually focussing on particular elements of their stock which are proving problematic" (Larkin 2000, p. 5). At the lowest level, local area option appraisal is probably the least well developed, but some work has been done by RSLs to develop approaches to the assessment or individual types of risk, with examples to be found in Larkin (2000, pp. 31–32), William Sutton Trust (2000, pp. 58–59) and Gruis and Thomas (2002).

Asset Management Process William Sutton Trust

The case description of the William Sutton Trust (2000) provides an example of a well-structured approach towards the strategic decision-making of a large association (15,700 homes) with a geographically spread housing stock. The process model that the Trust has devised with the assistance of HACAS Chapman Hendy is in three parts:

1) Estate prioritisation: based on a questionnaire completed by Trust employees working at area or estate level, all the estates were evaluated against a "sustainability index" and estates "at risk" were identified;
2) Estate assessment: the estates that had been selected in the previous stage were assessed on the basis of more detailed criteria and management strategies were developed for each estate;
3) Option appraisal: the options that had been determined in stage 2 were evaluated on the basis of financial costs and benefits and 'social' benefits in the form of the expected impact of measures in relation indicators used in the sustainability index (see William Sutton Trust, 2000 for a full description).

Estate Categorisation Bradford and Northern

Bradford and Northern have analysed their entire stock and placed the properties in one of three categories:

1) 'Core stock', which displays some or all of the following characteristics: good location, context and environment; healthy demand; low turnover; low number of voids; low rent loss; generates surpluses; exhibits a variety of property conditions. These properties are maintained and modernised with priorities identified through the estate management plans;
2) 'Redundant stock', which has some or all of the following characteristics: poor location, context and environment; no market envisaged; no longer meets housing need; low or no demand; high level of voids; high rent loss; high reactive maintenance spend; more likely to need rehabilitation; does not generate a surplus. These properties will be disposed of (presumably by sale or demolition), although some mothballing of properties where a comprehensive neighbourhood regeneration strategy is in place or envisaged. They are maintained to minimum legal standard where tenanted;
3) 'Maintained stock', which falls in neither of the above categories and requires more research until a firm view can be taken. They are maintained while marketing plans, further analysis and market monitoring are conducted to determine future directions (Larkin, 2000, p. 17).

Option Appraisal Process HACAS Chapman Hendy

HACAS Chapman Hendy with a Midlands based RSL have adopted a six step process towards option appraisal.

Step 1: Define the scheme or group of units to be assessed
The first step is simple but important: a definition of the stock to be included in the option appraisal. The aim is a boundary which makes sense as a 'housing market' description rather than an administrative unit. That is, it should have a meaning to those who live in the area, or who might wish to live there.

Step 2: Prepare a base financial forecast
This step produces a forecast of income and expenditure for the target stock. Taking a narrower view than business planning, financial appraisal is limited to a simple measure of yield based on Net Operating Income (NOI) in relation to value. NOI is defined as rent collected less maintenance expenditure and void management costs. Valuation can be capital value (existing use) or discounted capital value (EUV-SH).

Step 3: Assess the area and cluster the stock
The ambition is to prepare a description of the area, describing its current housing function and assembling available evidence about likely demographic trends, housing demand, social need and projected rent profiles. The need for clustering will depend on diversity. If the appraisal is of a single tower block or estate, then homogeneity means that clustering will not be required. However, if the neighbourhood is diverse, and the stock dispersed (which is often true of a traditional RSL operating in the older urban areas), then different street, or parts of streets may be perceived differently, and show different performance.

Step 4: Identify the options and agree success criteria
Relatively low NOI is usually the combined result of low income (high arrears, high void loss) and high costs (high void expenditure, exceptional management costs, and possibly high costs associated with the age or construction of the property). Identifying options is concerned with increasing income and reducing costs. The opportunities are considered from the perspective of both current and potential tenants. The aim is to agree the predicted effect of possible actions. For example, investment in improved standards might reduce void loss and management costs, increase satisfaction with the property and have a possible effect on cyclical costs and satisfaction with the area.

Step 5: Prepare option forecasts
The next step is to apply each option in turn to the base model to forecast the effect of either individual courses of action or the cumulative impact of two or more options applied together. The modelling problem is to deal with the uncertainty of individual impacts. For example, while the costs of redevelopment or disposal are relatively clear cut, management interventions are more difficult to predict. Though it is possible that additional capital investment will improve tenant satisfaction and reduce levels of turnover, the precise effect is impossible to predict. A range of possible outcomes is a much more satisfactory approach, and this suggests risk simulation as the preferred approach.

Step 6: Agree the preferred option, and define the implementation plan
With agreement on the preferred option the final step is to define the implementation plan. The option appraisal will already have made assumptions about targets, costs and timing for each component of the action plan (this being integral to risk modelling). This has to be translated into an action programme.

6.4 CONCLUSION

English local authorities and housing associations are under pressure to develop a more strategic approach towards asset management for various reasons. The mixed funding regime for RSLs introduced under the Housing Act 1988 exposed associations to financial risk and one of the implications was that it created a climate favourable to the idea of business planning. Local authorities followed more slowly, but the introduction of resource accounting began a process which has now developed into the requirement for business level option appraisal.

Despite these developments, the field of asset management is not yet well established in England, and there are many weaknesses to be tackled. The data being used has weaknesses, information systems are poorly integrated, and option appraisal is unsophisticated. There is little evidence in the literature that English social landlords in general have used asset management as a way of refocusing their business, although serious efforts have been made to develop a more systematic and comprehensive view of asset management by some front-runners.

Nevertheless, asset management in England has grown significantly in importance over the past ten years and is likely to develop further. Though the continuation of present trends is by no means certain, the current thrust of government policy is that local authorities will cease to be the direct providers of social housing, and that RSLs in their various forms will be responsible for its management under a common regulatory framework. This means that high level investment plans, to support the business planning process, will establish themselves as a standard feature of RSL management. The key relationship between investment and business plans has already introduced the concept of assets which produce a variable return, and one of the features of recent years has been an acceptance of the idea of a poorly performing sub-set of stock which needs to be identified and managed. Most encouraging of all, perhaps, is a recognition that asset management is not the exclusive preserve of housing maintenance. Portfolio performance is now a business issue, with lettings policies, rents, marketing and management strategies all having a role to play.

REFERENCES

Audit Commission (2003) A framework for the review of housing inspection and assessment: Draft for consultation, London, AC.

Barker, K. (2003) Review of housing supply: securing our future housing needs, London, HM Treasury, HMSO.

Brown, T. (2000) Needs versus choice in social housing allocation systems. Paper presented at the ENHR conference, Gävle.

Brown, T., Richardson, J. and Yates, N. (2002) Whose asset management?, Paper presented at the ENHR conference, Vienna.

Blackaby, B. (2000) Understanding local housing markets, Coventry, Chartered Institute of Housing.

DETR (1998) A new financial framework for local authority housing: resource accounting in the Housing Revenue Account, Consultation Paper, London, DETR.

DETR (2000a) Quality and choice: a decent home for all, London, DETR.

DETR (2000b) A new financial framework for local authority housing: guidance on business plans, London, DETR.

DTLR (2001) A new financial framework for local authority housing: resource accounting in the HRA-HRA subsidy and rent restructuring, London, DLTR.

DTLR (2002) A decent home: the revised definition and guidance for implementation, London, DTLR.

Gruis, V. and Thomas, A. (2002) Approaches to option appraisal in social housing investment: English and Dutch experiences compared, Paper presented at the ENHR Conference, Vienna.

Heijden, H. van der (2002) Social rented housing in Western Europe: developments and expectations, Urban Studies, 39, pp. 327–340.

HM Treasury (2003) Review of the delivery of the decent homes target for social housing PSA Plus Review, London, HMSO.

Housing Corporation (1999) Good practice guide to maintenance cost forecasting, London, Housing Corporation.

Housing Corporation (2002) The way forward: our approach to regulation, London, HC.

Keenan, P., Lowe, S. and Spencer, S. (1999) Housing abandonment in inner cities: the politics of low demand for housing, Housing Studies, 14, pp. 703–716.
Larkin, A. (2000) Asset management strategies; a review of asset management strategies of housing associations in England and social housing providers in Australia, London, Metropolitan Housing Trust/The Housing Corporation.
Malpass, P. (2001) The uneven development of social rented housing: explaining the historically marginal position of housing associations in Britain, Housing Studies, 16, pp. 225–242.
Murie, A. (2002) Changing demand, London, Housing Corporation.
NAO (2003) Improving social housing through transfer, London, National Audit Office.
Newey, M. (2002) Homes vs. life cycle costs; practical lessons in asset management from an English housing association, Paper presented at the ENHR conference, Vienna.
NFHA (1994) Stock condition surveys: a basic guide for housing associations, London, National Federation of Housing Associations.
NHF (1997) Reinvestment strategies: a good practice guide, London, National Housing Federation.
ODPM (2002) Allocation of accommodation: code of guidance for local housing authorities, London, ODPM.
ODPM (2003a) Sustainable communities: building for the future, London, ODPM.
ODPM (2003b) English house condition survey 2001, London, ODPM.
ODPM (2004) Making it happen: the Northern way, London, ODPM.
Pawson, M. and Mullins, D. (2003) Changing places, housing association policy and practice on nominations and lettings, Cambridge, Policy Press.
Walker, R.M. and Smith, R.S.G. (1999) Regulatory and organisational responses to restructured housing associations finance in England and Wales, Urban Studies, 36, pp. 737–754.
Walker, R.M. and Zon, F. van der (2000) Measuring the performance of social housing organisations in England and the Netherlands: a policy review and research agenda, Journal of Housing and the Built Environment, 15, pp. 183–194.
Walker, R. (2001) How to abolish public housing: Implications and lessons from public management reform, Housing Studies, 16, pp. 675–696.
Walker, B. and Marsh, A (2000), Rent setting policies in English local authorities 1998/99, London, DETR.
Walker, B. and Marsh, A. (2003) Setting the rents of social housing: the impact and implications of rent restructuring in England, Urban Studies, 40, pp. 2023–2047.
William Sutton Trust (2000) Stock investment and social change; a case study for a registered social landlord, Tring (Herts, UK), William Sutton Trust.

7 FRANCE[1]

FRÉDÉRIC BOUGRAIN

7.1 SOCIAL LANDLORDS IN FRANCE

In 1999, there were about 24.5 million dwellings in France (Ministère de l'Equipement, 2001). The social housing sector accounted for about 20% of the total housing stock (Table 7.1).

The social rented stock is mainly managed by two categories of social landlords: Municipal/departmental housing companies and private social housing companies (Table 7.2). In 2001, there were 201 housing companies owned by municipalities and 91 owned by 'départements'—regional administrations comparable to counties or provinces. These municipal or departmental housing companies are divided into two major groups: public HLM (Habitation à Loyer Modéré—low-cost-housing) organisations (Office Public d'Habitation à Loyer Modéré—OPHLM) and public development and construction organisations (Offices Publics d'Aménagement et de Construction—OPAC). Public HLM organisations are public administrative bodies, controlled by local authorities (municipalities, municipal associations or 'departments'). Their main characteristic is to operate on behalf of a local authority and to implement local authorities' social housing policy. They were created in 1912 (Ministry of Housing, 1997). Public development and construction organisations operate within a hybrid legal framework which mixes elements of public and private law. They operate also on behalf of a local authority. They were created in 1971.

Private social housing companies were created during the nineteenth century. There were 317 private companies in 2001 (UNHLM, 2001). Many of them have been integrated in large groups.[2] 119 private social housing companies are gathered in 32 groups and together they own 52% of the dwellings. The 10 largest companies own about 25% of the housing stock and manage more than 10,000 dwellings each. The largest group manages about 130,000 dwellings. In recent years, the share of private social housing

[1] A first draft of this paper benefited from the comments of Jean Carassus. I would like to express him all my deep gratitude.

[2] Under this scheme, one main housing company controls the capital of several others. Financial and human resource management are shared to benefit from economies of scale but each of the affiliates still manages its own assets.

Vincent Gruis and Nico Nieboer (eds.),
Asset Management in the Social Rented Sector, 119–139.

Table 7.1 Division of housing stock by tenure in France (1999)

	Owner-occupied	Private rented	Social rented (municipal, departmental and private social housing companies)	Social rented (other social landlords)
number of dwellings (in millions)	13,403	6,209	3,647	1,223
percentage of total stock	54.7%	25.4%	14.9%	5.0%

Source: Ministère de l'Equipement (2001).

Table 7.2 Social rented housing companies in figures (2001)

OPHLM	OPAC	Private social housing companies
47 departmental offices managing an average of 8,000 dwellings	44 departmental offices managing an average of 15,300 dwellings	316 private companies managing an average of 5,650 dwellings
148 municipal offices managing an average of 4,000 dwellings	53 municipal offices managing an average of 9,600 dwellings	
969,500 dwellings	1,182,700 dwellings	1,784,800 dwellings

Source: Fédération nationale des sociétés anonymes et fondations d'HLM (2002).

companies increased: they managed 1.2 million dwellings in 1993 and 1.78 million in 2001. Other social landlords can be divided into two main groups: Co-operative housing companies which are subject to the same legislation as municipal and private housing companies, but whose main role is not to rent dwellings but to favour owner-occupancy. They own about 14,000 dwellings; 250 public-private housing companies which are owned by municipalities and manage about 520,000 dwellings.

In this chapter, we confine ourselves to the private social housing companies because they are considered as being more competitive, better managed and more market-oriented than those owned by local authorities. They are not-for-profit organisations but governed by shareholders.[3] Private housing companies

[3]Before August 2003, one shareholder could not have more than 10 'voices'. Consequently, even a majority shareholder was not able to control a housing company by itself. This situation recently lead to a crisis opposing shareholders and managers. The majority shareholders asked the government to repeal the rules concerning the '10 voices'. The law enacted on the 1st of August 2003 modified the governance of housing companies. Now a majority shareholder will have the majority during general assembly, but to counterbalance this power, local authorities and representants of tenants will benefit together from one third of the voices in general

are usually sponsored by private businesses, but they may also be sponsored by public companies or local authorities. In this way, the organisations try to make sure they can provide housing for their employees. Since reforms of social housing in 1977, the private social housing companies have become the main producers of social housing. They are also involved in the construction of subsidized owner-occupancy dwellings for lower income groups (see also Boelhouwer (ed.), 1997).

7.2 NATIONAL CONTEXT FOR ASSET MANAGEMENT

Although the social housing companies are 'private' institutions, they are subject to regulation by the French government concerning for example rents, allocations and investments. However there have been some changes recently because of the new government policy towards asset management. In 2002, the government strongly encouraged social housing companies to develop a strategic management approach to optimise their scarce resources and to put an end to the misallocation of grants. In the future, social housing companies without a strategic approach will not be able to receive financial subsidies to build or refurbish their housing stock.

7.2.1 Regulations

General administrative regulations

The law which was enacted in 1922, stipulated that housing companies had to encourage the construction of decent and cheap houses to lower-income households. Since then, this social aim, "the right to accommodation", has not changed, but several decrees have modified the legal framework, concerning for example the creation of subsidised rent loans, personalised housing subsidies, the distribution of responsibility between local authorities (municipalities, "departments" and regions) and the State, the promotion of home-ownership, the relationships between tenants associations and housing companies and the transformation of offices into dwellings.

In 2000, a law was enacted to combat social exclusion. This act strongly modified the approach of housing companies. Its aim is to reduce the number of dwellings which have a poor quality and to stimulate the participation of housing companies in the urban renewal process. Nowadays, refurbishment or construction projects go beyond the competencies of a single housing company; projects have progressively become a joint initiative of housing companies

---

assembly. Housing companies have two years to modify their memorandum and articles to comply with this law.

Table 7.3 Monthly rents according to the location of the dwelling (in euros per m^2, 2001)

	Social rented sector	Private sector	Average
Urban areas with less than 20,000 inhabitants	3.5	5.0	4.3
Urban areas: between 20,000 and 100,000 inhabitants	3.5	5.3	4.3
Urban areas: between 100,000 and 2 millions inhabitants	3.8	6.5	5.2
Paris and its suburbs	4.5	10.9	7.2
France	3.9	7.0	5.4

Source: INSEE (January 2002).

and municipalities. The inhabitants and the intermediary organisations in the neighbourhood, such as schools, the police, welfare and religious organisations, can also be involved. This requires close collaboration between all these stakeholders and modifies the traditional approach of housing companies.

Rent setting

The rents are subject to regulations by the State. The level of the rents varies according to the financial supports schemes which have been granted to the housing companies for the construction of the dwellings. The housing companies are not free to increase the level of the rent. The aim of the regulation is to make social housing affordable for people with low incomes. Consequently the rents for the social rented sector are much lower than the private rents in big cities. Conversely, in rural areas, there is almost a balance between social and private sectors (Table 7.3).

Allocations

Allocations are also constrained by regulations. There are income ceilings for those who intend to rent dwellings in the social housing sector. The income ceilings are revised by the government every year and vary according to region and financial support scheme. For example, the income limits for dwellings located in Paris and built with PLI[4] are three times higher than the limits for dwellings located outside the suburbs of Paris and built with PLAI[5] (Table 7.4).

---

[4] PLI (Prêt Locatif Intermédiaire) are programmes dedicated to finance intermediary dwellings. They are financed by loans with low interest rates.

[5] PLAI (Prêt Locatif Aidé d'Insertion ou d'Intégration) are programmes benefiting from loans with low interest rates. Moreover, in this scheme, social housing companies are entitled to benefit from State subsidies, measuring up to 20% of the operation-costs.

Table 7.4 Annual income limits for access to the social rented sector (January 9th, 2003)

	Taxable income (in euros)			
	Paris and the surrounding municipalities		Other regions	
	PLAI	PLI	PLAI	PLI
One person	11,648	31,766	10,126	27,619
Two people	18,989	47,475	14,753	36,882
Three people or one person with one dependant person	24,894	62,237	17,740	44,354
Four people or one single with two dependant people	27,246	74,304	19,740	53,544
Five people or one with three dependant people	32,415	88,407	23,096	62,987

Source: http://www.logement.equipement.gouv.fr.

In 2000, about 11% of people who lived in social dwellings had an income above the limits. However, households with increasing financial resources are not required to move and leave their dwelling. They just have to pay a supplement to compensate for their better financial situation. Nevertheless, tenants who live in the social rented sector are, relatively speaking, poorer than 30 years ago[6]. In the seventies, the average income was approximately the same in the private and the social sectors. In 1996, the average income was 25% lower for tenants of the social rented sector. In 2001, 73% of the households who moved in a social dwelling, had resources lower than 60% of the ceilings.

Sales

As part of privatisation policies, housing companies have been encouraged to sell some dwellings to their tenants. In the past, tenants initiated the sales, but since 1983, housing companies initiate the sales. The legislator expected that housing companies would sell about 30,000 dwellings per year. However, the number of sales has never reached such a high peak (see section 7.3.3).

Maintenance and renewal

The building and housing code states the minimum maintenance requirements. Maintenance is required to ensure safety standards for elevators, heating

[6] Between 1975–1985, many households moved out of the social rented sector and bought individual houses by benefiting from governmental subsidies.

systems, gas appliances, collective gas-controlled mechanical ventilation systems, garage doors, detection systems, fire protection, etc. A law enacted in 1989 lays down the responsibilities of landlords and tenants. Tenants pay for ongoing maintenance through service charges that are not connected to their income level but to the characteristics of their dwelling.[7]

7.2.2 Supervision

All social housing companies are monitored by one organisation: MIILOS (Mission d'Inspection Interministérielle du Logement Social) which is related to the Minister of Housing and Minister of Finance. About 200 controls are made every year. These controls concern accounting, financial, social, technical and administrative aspects. MILOS can impose sanctions if the rules have not been respected. These sanctions are either pecuniary or organisational. If the infringement is strong, the board of directors may be dismissed. Due to its supervision activities, MIILOS knows how housing companies are being managed. Thus it is entitled to propose new solutions to improve the management of the social rented sector.

7.2.3 Financial Support

The French housing sector is still heavily subsidised compared with some European counterparts where governments have diminished their financial support. Both housing companies and households are the main beneficiaries of the financial support system. Housing companies benefit from low interest loans to finance:

– the purchase of land and the building of new houses[8];
– the purchase and the refurbishment of existing dwellings (in this case, the costs of the refurbishment has to represent at least 25% of the purchase price);
– the transformation of offices into dwellings.

Some subsidies for refurbishment projects amount up to 10% of the total cost of the operation (within a limit of 13,000 euros). Moreover, the value-added

[7] For example, charges for elevators are higher for tenants living on the 7th floor than on the second one.

[8] On average, State subsidies represent 13.5% of the costs of a building project; 5.5% come from the associations managing the employers' contribution to the construction effort and 1.6% stems from local authorities. The level of housing companies' own resources is only 5.4%. Thus 74% of the costs of new building projects are financed by loans (Les Echos, 12 June 2003).

Table 7.5 Rent limits for a dwelling of 80 m^2 dedicated to a household with two children (in euros per month—in January 2003)

Nature of the financial support (from the most subsidised to the less)	Paris and the surrounding municipalities	Parisian area (without Paris and the surrounding municipalities)	Other regions
PLAI (20% of subsidies)	371	306	283
PLUS (5% for new houses and 10% for renewal)	417	344	320
PLS	625	516	479
PLI (no subsidies and still low interest rate).	792	619	575

Source: http://www.logement.equipement.gouv.fr.

tax concerning all renovation and rehabilitation works has recently decreased from 20.6% to 5.5%.

The building programmes that do not benefit from subsidies and from tax refunds are those that are devoted to households whose resources are too high to have access to the less expensive dwellings but too low to rent a dwelling in the private sector (Table 5).

There are ongoing changes in the financial framework. The government has reduced direct financial subsidies and modified the financial institutional framework. In August 2003 it has been decided to create a National Agency dedicated to urban renewal. Between 2004 and 2008, this Agency will receive 1.2 billions euros every year. The aim is not to sustain only housing companies but mainly to promote urban renewal in neighbourhood suffering from social problems (ZUS—Zone Urbaine Sensible—literally, "Sensitive Urban Area"). The Agency will first focus on 160–170 very poor neighbourhoods.

The agency will coordinate local urban projects and the financial allocation of funds at neighbourhood-level. The financial means of the agency will be devoted to refurbishment projects, demolition, construction of social housing, urban rehabilitation or creation of schools. Its participation in each project will depend on the wealth of the municipality. Ten projects have already been validated. These projects will represent an investment of 1.14 billion euros (the contribution of the Agency will equal 380 million euros). They will concern the construction of 5,520 dwellings and the demolition of 6,925 (Les Echos, November 2003). To comply with this new institutional framework, housing companies will have to intensify their collaboration with the other stakeholders in the neighbourhoods (local authorities, financial institutions usually involved in social housing, tenant associations, other housing companies).

There is also a possibility for tenants to receive a housing allowance. The level of the Household Housing Aid ("APL"), designed to reduce the direct costs paid by the tenants, depends on the taxable income of the household. In 2001, 53% of the tenants who lived in the social rented sector, received APL (Les Offices de l'Habitat, 2003).

Private social housing companies are in a relatively better financial position than other housing companies. In 2001, half of them had a ratio of self-financing over rental income higher than 10 percent[9] (only 30% of municipal social housing companies are at the same level). About 10% have a negative ratio and can be considered in a bad financial position. Most of them own less than one thousand dwellings (Fédération Nationale des Sociétés Anonymes et Fondations d'HLM, 2002).

7.2.4 Housing Market

The social housing market in France is not balanced, from a quantitative nor from a qualitative point of view. In most regions, there is no equilibrium between supply and demand. Most of the housing stock is located either in Paris and its suburbs or around large cities. This results from the industrial growth after World War II and the rural depopulation. It is also around the large and most dynamics cities that the largest housing shortages occur. Some figures: In 1996, on average, 8% of the tenants had to wait for 3 years for their dwelling; 40% waited less than 3 months. Of the people living in Paris and its suburbs, 58% waited less than one year, while 81% of the demands which concerned the rest of the country were satisfied within one year. This scarcity is becoming worst and worst every year, due to several causes:

- Demographic and sociological trends lead to an increase in the number of single households. Elderly people stay longer and longer in their dwellings; Divorces in big cities are more numerous than in the past; Young people prefer to leave alone before getting married[10]. In 1999, they were 7.4 millions single households. 43.5% lived in the centre of large cities;

[9] This ratio is calculated as the net resources (resources minus charges such as taxes, wages, maintenance) dividied by the rental income.

[10] In 1982, 57% of the people aged 25 lived with a partner. In 1999, this was only 37% (INSEE, December 2002).

- Then number of demolitions tends to increase. The Ministry of Housing would like to see about 40,000 demolitions per years for the next five years and the number of demolitions had already increased to about 8,000 in 2002 (see Table 7.6 for previous years). Moreover, mostly high-rise apartment blocks are being demolished, while new building projects are limited in size. These building with a maximum of four stairs are more attractive and they are easier to integrate in residential neighbourhoods.
- Constructions have strongly decreased due to the reduction of direct financial support and the fact that acquisition of new building land in large cities is becoming more and more difficult and expensive.
- On average, the quality of social dwellings is better and the size larger than those rented in the private sector (all of them have central heating, toilets and sanitary facilities).

7.3 ASSET MANAGEMENT PRACTICE

In the following sections we will discuss asset management in the French social rented sector. Our analysis draws from approaches of some large private housing companies. This category of landlords seems to have relatively dynamic approach towards their asset management. Moreover, some of these companies have their assets located in several regions, which enlarges the need for a strategic approach. However, due to the limited number of cases, it is not possible to consider their approach as representative for all housing companies.

7.3.1 Rent Setting

As stated in section 7.2.1, social housing companies are not free to set their rent. In 1999, they signed an agreement with the Ministry of Housing. The government diminished the interest rates of 'social' loans that housing companies had contracted in return for a freezing of the rents for 2 years. The example below illustrates this situation.

Example of a Private Social Housing Company's Evolution of Rental Income 1999–2000

As a result of the act described above, in 2000 and 2001, the only way for housing companies to increase their rental-income was by refurbishments and acquiring new properties. As we can see in the table below, the rental income of a large housing company (over 40,000 dwellings) has increased mainly due to the construction and acquisitions of new estates.

Resources	1999 (×1,000 €)	2000 (×1.000 €)	Difference 1999–2000	% of increase
Rents from the housing stock	112,578.14	112,743.71	165.57	0.15
Refurbishments	4,785.18	5,001.83	216.65	4.53
New dwellings	160.39	1,304.16	1,143.77	
Acquisitions	0	2,769.02	2,769.02	
Parking rents	5,025.61	5,397.47	371.86	7.40
Other renting resources	316.97	318.80	1.83	0.58
Supplement of rent paid by tenants with taxable income above the resources limits	861.87	949.99	88.12	10.22
Rents from collective houses	2,817.50	2,830.16	12.65	0.45
Rents from retail units	1,988.56	2,205.52	216.95	10.91
Total resources	128,534.23	133,520.66	4,986.43	3.88

Source: Logement Français, 2001.

7.3.2 Allocations

Dwellings are usually allocated on the bases of the financial resources of the households (see section 7.2.1). However, housing companies may reserve some of their dwellings for local authorities in return for their financial guarantees. Furthermore, part of the dwellings is reserved for employees working in companies that finance social housing through the "1% housing" ("1% logement")[11]. Those employees also have to conform with income criteria to be eligible for a dwelling.

[11] The "1% logement" is paid by any private company employing more than 10 people. It was initiated in 1943 by managers from the North of France who wanted to finance partially the housing of their employees. It became mandatory in 1953 to contribute to the construction effort. The rate, initially set at 1% of the total wage bill of private companies, has been set up at 0.95% since 1992. Private companies usually give their contribution to registered intermediary organisations that collect these funds. As a compensation, some dwellings are reserved for their employees. Thanks to this system, registered intermediary organisations became the main shareholders of private housing companies.

In 2001, 726 million euros have been dedicated to investments in social dwellings. 1,698 millions benefited to households. 518 millions financed loans with zero interest rates.

'Innovative Marketing' of Problematic Estates

Within the restrictions of the allocation-rules, some housing companies have developed new methods to attract tenants to problematic estates. One large private social housing company, owning about 50,000 dwellings, has replaced the 'traditional' rental department with a marketing department in estates characterised by social difficulties or significant vacancies. The aim is to better match the needs of the tenants with the characteristics of the dwellings. Moreover, individual visits have been replaced by collective visits to increase the feeling of scarcity among the potential tenants.

7.3.3 Sales

Although the government expected that housing companies would sell about 30,000 dwellings per year, sales of dwellings to tenants have been far less. As we can see in Table 7.6 the number of sales have remained at a level of around 5,000 dwellings a year. Several reasons may explain this situation:

- Financial incentives for buying are limited. The tenants who want to buy their flat have to pay the market price. They can only benefit from loans with a low interest rate;
- The financial situation of tenants is worse than thirty years ago. Social housing is not anymore a 'transitional' tenure as it was during the seventies. The middle class has already moved out of the social rented sector and bought dwellings by benefiting from governmental subsidies;
- Most of the housing stock (87%) consists of multi-family blocks. In Paris and its suburbs one counts only 3% of single-family dwellings, while tenants usually prefer to buy single-family dwellings. Several new dwellings consist of single-family dwellings, but housing companies are

Table 7.6 Development of the social housing stock in France (number of dwellings, 1995–2001)

	1995	1996	1997	1998	1999	2000	2001
New	54,900	42,700	43,800	37,700	37,300	32,700	33,100
Buy and renewal	8,000	8,500	6,200	6,100	6,500	7,200	9,000
Demolition	4,600	3,100	2,900	4,000	5,600	4,400	5,200
Sales	3,400	4,300	5,200	4,500	6,100	4,400	3,900
General evolution	54,900	43,800	41,900	35,300	32,100	31,000	33,000

Source: M. Amzallag and C. Taffin (2003)

not willing to sell the 'jewels' of their portfolio which are easy to rent as well;

- Housing companies do not like to manage joint-ownership buildings, because of legal and social difficulties.[12]

However, sales might increase in the coming years due to the budgetary constraints posed by the intended reduction of financial government support. The Ministry has suggested that housing companies could sell part of their housing stock to improve their financial position and to compensate for the reduction of subsidies. The government considers that the profit realised from one sale might equal in value the subsidies that housing companies used to receive for the construction of three new dwellings. But, most housing companies are still reluctant to sell dwellings which have been paid off and are very profitable (Le Moniteur, p. 70, 13 June 2003). On the other hand, it is recognized that sales represent several advantages for housing companies:

- They increase the differentiation of population within one estate. This helps housing companies to comply with the arguments of the government who looks for a mix between owners and tenants;
- They improve the financial position of housing companies that make capital gains by selling dwellings. But their financial situation is not so bad to urge them to sell;
- They are part of strategic asset management. Housing companies with a portfolio scattered in several municipalities have an incentive to sell their dwellings located in cities where they have a very small market share.

7.3.4 Maintenance and Renewal

About 60% of the social housing stock dates from before 1975. Many of these dwellings suffer have a mediocre technical quality and a poor market position. Maintenance and renewal policies have been considered for several years as one of the main weaknesses of housing companies. Several factors can be brought up to explain the poor past performances in this field.

- During the nineties, maintenance expenditures have decreased, due to the relatively poor financial position of most housing companies;

[12] To avoid potential conflicts, one housing company decided to refurbish all the dwellings belonging to the same block before selling part of them to the tenants. It has also developed a financial arrangement in which the landlord buys back the dwelling when tenants encounter financial difficulties.

- Cultural elements play a role as well. Allen and Bonetti (1998, p. 10) state: "Historically, social housing companies based their culture on the construction and the development of real estate investments that represented a noble, status-enhancing function". In this view, refurbishment and maintenance are regarded less important;
- The role of the employees dealing with the maintenance of the housing stock has changed. Their role is not anymore to decide or to develop new technical solutions but to monitor subcontractors, to plan and manage maintenance activities. However, most of them had a technical background and did not accept or were not qualified for the move from an operational to a functional activity (Bonetti and al., 1998);
- Mid and long-term maintenance was (and is still) often a source of misunderstanding between managers. The maintenance managers, who operated on a central level, relied on technical criteria to plan long term maintenance, while the managers of the estates carried out the repairs required by tenants. The latter argued that the decisions taken by the maintenance manager were based on technical criteria and lead to tenants' dissatisfaction;
- Most rehabilitations were technical and aimed at improving the energy efficiency of the buildings. Outdoor areas and the neighbourhoods were neglected and most rehabilitations did not fit with tenants needs;
- Housing companies tended to carry out substantial renewals when considerable financial subsidies were available;
- The refurbishments could not compensate the lacks of the original design.

The mitigated effects of past policies has lead some housing companies to modify their approach towards mid- and long-term maintenance and renewal. This can be seen in various ways.

- Housing companies begin to understand that the value of their assets is not only connected to the technical characteristics and performances of the buildings. Urban, social and economic factors have been integrated to the analysis. Working on the neighbourhoods in co-operations with local authorities has become progressively the rule;
- The management of refurbishment projects has been completely rethought. Project managers have been hired to carry out rehabilitations. Operational teams involving economists, technicians, managers, architects, town planners, sociologists are at the core of these projects;
- The casualness of applying standardised maintenance methods disappears. Housing companies tend to develop a step by step approach to take into account tenants needs and to learn from one project to one other.

Asset Management and Urban Renewal: Example of a Municipal Housing Company (Bonetti et al., 1998)

By implementing strategic asset management, a municipal housing company aimed at improving the allocation of scarce financial resources dedicated to maintenance and renewal. Its new approach articulated planned, responsive and cyclical maintenance requirements by taking into account technical, commercial and social criteria. Previously, maintenance was organised according to the technical shape of the buildings and the subsidies received from the government. The new method involved both financial and technical managers. It leads the housing company to classify its estates according to:

- their urban location, the quality of the neighbourhood and the services provided;
- their technical quality;
- the social characteristics of the tenants;
- the maintenance approach.

This classification helps the landlord to plan its maintenance and renewal strategy according to its social objectives, market constraints and the evolution of the environment.

7.3.5 Decision-Making Process: Considerations, Methods and Strategies

One of the aims of strategic asset management is to analyse and anticipate technical, commercial and economic risks. It enables to adopt a better strategy in front of the evolution of the environment and to end unsuccessful refurbishments. By prioritising their investments, housing companies avoid ineffective allocation of funds (for example by allocating funds to the estate managers who have the best image within the company—a kind of "populism").

As stated in the previous section, in the past the refurbishment policy subsidised by the government has been not very effective. This lack of effectiveness was also connected to the absence of asset management, which is not conducted in a systematic manner among housing companies yet. Several of them have not formalised their methods and their procedures have remained tacit. However, some housing companies have started in the early nineties to implement strategic asset management, among others to improve the quality of the global services provided to the tenants (the quality of the dwelling, outdoor areas and neighbourhood services). Several companies have also improved their strategic management methods when they increased their size through horizontal integration. Recently, the government urged housing companies to implement asset management in order to increase the efficiency of its urban policy. In the future, those which will not develop a strategic approach will not be able to receive financial subsidies to build or refurbish their housing stock. The government considers that strategic asset management will "help housing companies to think about the attractiveness of their asset, the social occupancy of the dwellings and to choose between small/large maintenance investment, rehabilitation, building new dwellings, strong refurbishment, partial or total demolition, sell to the tenants or to another company" (Ministère de l'Equipement, 2003). Asset management is also necessary to deal with the evolutions of the competitive environment:

- Some tenants are attracted by parts of the private sector that may provide dwellings with better services and similar rents (Bat'im Club, 2001);
- The increase of the unemployment rate leads to an increasing number of tenants with low financial resources;
- Some 'distressed' areas have become unattractive;
- The financial support of the government changed. Part of the subsidies decreased and were compensated by lowering the value-added tax on rehabilitation works (Bonetti at al., 1998).

For housing companies that have implemented strategic asset management, reorganisation was the cornerstone of their strategy:

- Collaboration between technical, financial, commercial and social departments has intensified;
- Co-operation with tenants and local authorities has been encouraged;
- Many organisations have been decentralised to facilitate the relationships with local partners.

These developments cause housing companies to switch from a technical approach towards a professional services attitude, thus strengthening the role of estate managers and improving the services provided to the tenants (Allen and Bonetti, 1998).

Environmental issues are not integrated in asset management yet. They are not as crucial for the housing company as social and economic aspects of sustainability. Most of the investments aimed at improving energy efficiency and water savings are conducted in order to reduce service charges and maintenance costs paid by the tenants. In the field of waste management, air conditioning, energy and water consumption, housing companies follow the legislation enacted by French public authorities. French housing companies have only just discovered the concept of sustainable building and their

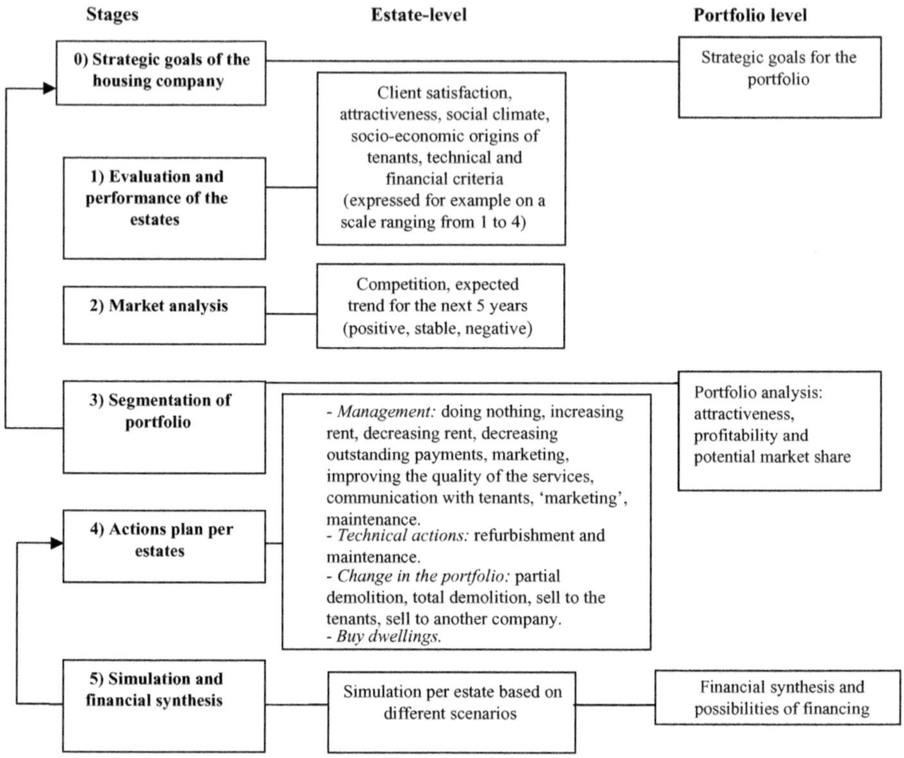

Figure 7.1 The strategic asset management process
Source: Carassus (1999).

environmental policy is still in their infancy. Subsidies and tax allowances which are delivered by the Agency for the Environment and for Energy Control (ADEME), are the major catalyst for changes in this area.

The decision-making process

The housing companies that have developed a more systematic approach towards their asset management follow a more or less similar procedure. Housing companies establish a link between their strengths and weaknesses and the characteristics of supply and demand for housing. Market analyses are based on several quantitative methods (annual survey of client satisfaction, market performance index, calculation models . . .). These analyses help housing companies to segment their stock. The examination of the performance of the portfolio of estates helps housing companies to establish financial plan and to examine the financial consequences of different investment scenarios. The synthesis will lead to actions plan per estate (see Figure 7.1).

Although strategic asset management is not widespread in the social housing sector, more professional approaches and techniques are being applied more and more every year. Particularly the large housing companies, managing dwellings in several cities, have improved their methods in the last five years. A first step of most housing companies is to carry out, on a yearly basis, large market inquiries to appreciate tenants' needs and satisfaction. These qualitative enquiries combined with quantitative figures such as vacancy rates, turnover, help the housing companies to classify their dwellings. Some housing companies have also integrated elements concerning the quality of the neighbourhoods in their portfolio analysis. From there, they define strategies for their estates. Most of these analyses were inspired by the Boston Consultancy Group approach. However, risk analysis (tests of strategies against different scenarios such as different interest /inflation rates levels) is not at the core of the new approaches.

Example of a Portfolio Analysis (Logement Français, 2001)

The renewal of the urban policy lead a private social housing company (owning over 40,000 dwellings) to classify its estates according to their strengths and weaknesses. Four categories have been identified:

1) Estates located in the periphery of large cities, which are highly demanded, close to transportation networks and offer all urban services (schools, banks, retails. . .). Households who rent these dwellings have usually encountered difficulties to have (immediate) access to the private sector. They are also expected to move to a different place sooner or later. For example, in the estates located in Aulnay-sous-Bois, for every dwelling which is vacated, ten demands are registered. Despite this situation, the housing company will not stop carrying out substantial renewals to keep these dwellings in a healthy market-position. From 2001 till 2010, a budget of 76,22 millions euros will be devoted to the 4500 dwellings located in Aulnay and to outdoor areas;
2) Estates located in areas away from local services such as banks or grocery stores. However, most of the inhabitants are attached to their neighbourhoods and are not ready to move. To improve this situation the housing company has invested 10 millions euros for three years to improve the quality of the neighbourhood by providing new services, to maintain the dwellings according to the tenants' needs, and to create a decentralised management unit to improve the interactions with the local environment (the tenants, the local authorities. . .) and to provide rapid solutions to arising problems;

3) Estates located in neighbourhoods characterised by economic depression, but. nothing indicates (yet) that this situation will last. The strategy of the housing company is to adjust temporarily its supply to the demand by stopping letting dwellings in some blocks and concentrate the tenants in other blocks in the neighbourhood. Thus, the landlords is able to save time and money on the empty blocks. On the longer run, the blocks are either demolished or, if the economic depression does not last, rented again.
4) The fourth category of estates, characterised by a high vacancy rate, will be partly demolished. The aim is to reduce the supply and to use the building land to propose attractive individual houses.

Strategic situation	Location	Number of dwellings	Vacancy rate in 2000
High perspective	Sartrouville	1,647	2.2%
	Goussainville	1,337	1.3%
	Aulnay-sous-Bois	4,428	1.7%
Poor location but strong attractiveness	Corbeil Les Tarterêts	1,737	4.4%
Estates hurt by economic depression	Creil	1,283	2.9%[13]
Poor future	Dammarie-les-Lys	1,709	6.6%
	Mantes	933	6.6%

Property exchange

Through the exchange of assets with other housing companies, the aim of some housing companies is to strengthen their location in some municipalities and to disinvest from others. By developing such a policy, the best managed housing companies intend both to reduce their structural costs and to improve the efficiency of the relationships with local authorities.

This policy has also been stimulated by the government which has recently established a new urban policy. It concerns 162 urban areas that have been qualified as "sensitive". In these areas, housing companies have the opportunity to receive more subsidies when they launch programmes aiming at improving the quality of the estates and their surroundings. To do so, housing companies are encouraged to co-operate with local authorities. In order to

[13] This vacancy rate does not include the blocks that are being left vacant intendedly as part of the landlord's strategy.

facilitate the implementation of such a policy, housing companies can either co-operate together to supply new services or exchange their assets.

Examples of Strategic Property Exchanges between Three Housing Companies

In 1999, housing company A bought from B 612 dwellings located in the suburbs of Paris. After this transaction, its housing stock in those cities reaches respectively 1860 and 663 dwellings. This policy has helped A to diversify its housing stock and to master the costs attached to its neighbourhood policy (in estates characterised by social problems, it has set up a decentralised organisation by strengthening the role of the local management of the estates).

Conversely, housing company A transferred to B 344 dwellings located in three different cities that were far away from its business management units. The market position of the housing company in those cities was very bad compared with its competitors. Consequently it was not profitable to develop partnerships with local authorities. Housing company A also sold 183 apartments located outside the suburbs of Paris that induced high management costs.

In 2000, housing company A bought from C 293 dwellings located in three cities, and sold 118 dwellings. The sales concerned dwellings located in two cities without growth prospective and with limited relationships with local authorities. Conversely, the apartments that were bought, increased the size of the estates, reduced relatively the overheads, strengthen the relations with the municipalities and improve the supply of the housing company.

Tenant involvement

For the last few years, French housing companies have tried to involve tenants in activities that directly concern them. Most of them are aware that there is a need to increase tenants participation. In the past, they just gave information to the tenants. There was no real discussion. Some tried to involve them in some of their currents activities such as maintenance of the dwellings by developing do-it-yourself tools, but usually tenants do not take part in the management of the housing company. However, the influence of the tenants might change at the estate-level. The incentive was provided by new legislation, which stipulated that tenants should participate in the management of the local estates. For example, one very large housing company arranged financial support for the improvement of tenant participation. It decided to dedicate one euro per dwelling per year to the running of the tenants' associations, and tenants representatives will receive the same amount of money for training on matters such as the management and the financial governance of a housing company.

7.4 CONCLUSION

In France, strategic asset management in the social housing sector has been not as widespread as in the private sector. Social housing companies are non-profit organisation which have not been market-oriented for a long period. However, during the nineties, there have been initiatives to develop a more sophisticated approach towards strategic asset management. Furthermore, several events have recently encouraged housing companies to modify their practices, among which:

- The government is strongly encouraging social housing companies to develop strategic asset management and, in the future, those which will not develop a strategic approach will not be able to receive financial subsidies to build or refurbish their housing stock;
- Several companies have to improve their approach because of horizontal integration;
- The performances registered by past refurbishment policies have been very poor;
- Competition from certain parts of the private sector.

It is difficult to classify the approach of the housing companies according to the typology that has been described in the introduction of this book. The diversity among housing companies is quite strong. Some are market-oriented and proactive while others react after potential problems have become reality. Because of the developments described above, there is a tendency to adopt a more market-oriented approach (particularly among the housing companies managing more than 10,000 dwellings), to develop asset management in a more systematic manner, to leave the old informal methods of decision-making and to become more proactive. However, it is too early to examine the results of this change and to know if this new attitude also leads to a better costumers' satisfaction.

REFERENCES

Allen B. and Bonetti M. (1998) Decentralisation of social housing companies in France : an on going process, Paris, CSTB.

Amzallag M. and Taffin C. (2003) Le logement social (Social housing), Paris, LGDJ.
Bat'im Club (2001) Stratégie patrimoniale.
Boelhouwer, P. (ed.) (1997) Financing the social rented sector in Western Europe, Delft, Delft University Press.
Bonetti M., Salagnac, J.L. and Weckstein, M. (1998) Les stratégies d'investissement et d'entretien en continu du patrimoine, Paris, PUCA—CSTB.
Carassus J. (1999) Analyse patrimoniale simplifiée, complémentaire d'une analyse financière d'un organisme d'HLM, Caisse des Dépôts et Consignations.
Echos (Les) (2003), L'Agence de renovation urbaine est prête à ouvrir son guichet, 17 November, p. 5.
Echos (Les) (2003) HLM : le chantier de la réforme, 12 June, pp. 47–48.
Fédération Nationale des Sociétés Anonymes et Fondations d'HLM (2002) Les entreprises sociales pour l'habitat – analyses et statistiques.
Fédération Nationale des Sociétés Anonymes et Fondations d'HLM (2001) Les SA d'Hlm partenaires dans la décentralisation—rapport d'orientation.
Figaro (Le) (2003) Jeudi 15 Mai, p. 8.
INSEE (2002) La demande potentielle de logements.
Logement français (2001) Rapport de gestion 2000.
Moniteur (Le) (2003) Construire des HLM: un casse-tête pour les organismes, 13 June, pp. 69–71.
Ministry of Infrastructure, Transport and Housing (1997) Housing policy in France, Housing and Building Department.
Ministry of Infrastructure, Transport and Housing (2001) Comptes du logement—Rapport à la Commission des comptes du logement, DGUHC—DAEI.
Ministry of Infrastructure, Transport, Housing, Tourism and the Sea (2003), Circulaire n°2003-21-UHC/IUH2/6 du 21 mars 2003.

8 GERMANY

JOACHIM BRECH

8.1 SOCIAL LANDLORDS GERMANY

Germany is traditionally a country with a relative high rental quota in the housing sector. Almost 60% of all households are tenants. In 1998, in West Germany, 43% of all homes were privately owned; in East Germany, 31% (Source: Statistisches Bundesamt).

In Germany, social housing is not limited to developer-builders with special legal status. Principally, any organisation can be active in the public housing sector, as long as it conforms to certain "requirements", such as communal occupation stipulation; rent control and limits, dependent upon the category of public financial backing, degree of floor-plan quality (e.g. room size) etc. In the end, low cost dwellings have been built almost exclusively by so-called communal housing companies and co-operatives.

The public supported "sozialer Wohnungsbau" or social housing is conceived as residential development for a wide spectrum of people. This concept has not been changed. Public housing programs focus on middle-income households, which are unable to obtain "acceptable" housing on the free market. Here objects or projects are supported, not persons. Regarding the lower income households, policy is based on the 'trickle-down' principle: in this case the greater availability of cheap, older housing through new residential construction. For households with lowest incomes rental assistance (Wohngeld) is available, thus, direct personal assistance.

Management policies towards housing developers are best understood in the history of their founding and growth. The social functions which the development companies assume is differentiated according to their influence and history. The direction of a developer's policies is more dependent on its goals and tradition than on a strict categorical entrepreneural form. However, all developers together continue to perform an important function in the lower segment of the market. Today's public assisted housing in Germany is represented by the following developer types:

- former quasi-public development companies, including communal housing companies;
- co-operatives;

Vincent Gruis and Nico Nieboer (eds.),
Asset Management in the Social Rented Sector, 141–160.

Table 8.1 Housing in Germany by tenure (1998) (× 1,000)

	Total (× 1,000)	Rental units (× 1000)	Owner-occupied units (× 1,000)	% of owner-occupation
States of former Federal Republic	27,487	15,652	11,845	43,1%
New states and East Berlin	6,331	4,357	1,975	31,2%

Source: Statistisches Bundesamt, vhw 3/June-Juli, 2002.

– private housing developers;
– national and international investors.

The latter group is not involved with new housing construction, as far as this still exists. Such investors purchase mostly larger, existing residential complexes. Sellers are insurance companies, the German Railroad, German postal service, industrial enterprises and, to a certain extent, communities.

Former quasi-public development companies
Following World War II, a so-called "public housing economy" was developed. This sector of housing development enjoyed privileged tax status, according to the "Wohnungsgemeinnützigkeitsgesetz" or Public-Use Housing Law. Companies belonging to this sector are owned by the following entities: communities and states; state-owned banks; former public institutions such as railroads, postal service; churches; trade-unions; insurance and private companies with employee housing. After the repeal of the Public Housing Law such companies became de jure housing developers, operating on the free market. However, to a large extent their housing is still public assisted, and thus subject to state regulation. The above named companies administer approximately 4.8 million dwelling units.

Table 8.2 Percentage distribution of housing ownership in Germany in 1990, after reunification

	Western Germany	Eastern Germany
Communities	ca. 3%	42%
Housing companies and co-operatives	13%	17%
Private ownership (small home-owners, small investors, insurance companies, private housing companies)	ca. 84%	42%

Source: DIW-Wochenberichte 42/1990

The communal housing companies comprise an important part of the developers formerly belonging to the "Public Housing Economy". Numerous municipalities still possess housing authorities of significant stature. These institutions are regarded as irrefutable parts of communal social policies, guaranteeing acceptable housing for underpriveleged social groups.

Size and structure of the former quasi-public companies vary considerably, from developers with a few hundred apartments, to companies with 50,000 and more housing units. In part these businesses remain "genuine" housing developers. However many have turned into extra-regional corporations.

Co-operatives

The more than 2,000 housing co-operatives, some dating back to the end of the 19th century, administer approximatly 2.2 million apartments, circa 10% of the total number of rental units.

Co-operatives, from the nature of their situation, limit their scope of activities. Their main principle remains: construction for specific groups, for example: railroad-, postal-service-, communal civil-servant-employees etc. Characteristic of such co-operatives is not only tenant membership and, thus, co-ownership in the organisation, but also administrative efficiency and zero-growth rate. Today these cooperatives assure low-cost and well administered housing for a limited, but not needy clientel. The administration of larger co-operatives differs little from that of private housing developers, notwithstanding the co-operatives' use of self-administration.

Differences exist between 'traditional' co-operatives, some existing for more than one hundred years, and 'young' co-operatives, from which many were founded in the 1980's. Amongst the traditional co-operatives, some own up to 7,000 dwelling units. To the extent their apartments were built after World War II, they are still subject to state regulations. As a tax-exempt company, these co-operatives are free to construct non-regulated housing. However, in order to retain their tax-exempt status, they are not allowed to sell apartments at a profit. The young co-operatives are mostly very small, often owning very few units. This type of co-operative movement has taken form in the last 20 years, as the traditional housing developers and co-operatives ignored the demands of some segments of the housing market. The legal status of a co-operative provides the much sought advangtages of democratic business forms, with a proportionately high use of self-administration.

Private housing developers

As mentioned above, any company can theoretically participate in public housing sector construction. Many developers have built low-cost housing with the use of capital obtained from public funding. Increasingly, national and international investment groups are purchasing large quantities of housing

Table 8.3 Some characteristics of housing developers in Germany (1998)

Number of housing developers (former quasi-public and co-operatives, excluding private developers, even those owning public-assisted housing)	3,176
Number of rental units owned by housing developers and co-operatives	6,887,400
Number of employees housing developers and co-operatives	65,000

Source: GdW

projects from communities, the state, state-owned enterprises (railroad, post and telephone services).

Information concerning the number of housing units owned by housing developers reveal little concerning social ramifications, as the companies possess different types of housing in their respective portfolios. Of interest is, for how many apartments and during which period of time regulations are valid (rent-control, communal tenant stipulation). The longevity of present limitations is dependent upon the type of regulation. Thus, absolute figures are irrelevant. In addition, once subsidised apartments which should no longer be regulated, remaining only subject to the normal limitations of comparable rentals (Mietspiegel), are still, for political reasons, regulated.

Evidently, these housing developers and co-operatives represent an economically relevant force, as evidenced in the quantity of dwelling units in their possession (almost 6.9 million) and the number of employees involved with planning and administration.

8.2 NATIONAL CONTEXT FOR ASSET MANAGEMENT

8.2.1 Regulations

General administrative regulations

The most important of recent developments in German housing policy is the return to more market-oriented economics (1980–1989). Worth remarking is the revoking of the Public-Use Housing Law ("Wohnungsgemeinnützigkeitsgesetz") in 1984. The resulting legal status of the former quasi-public development companies, as "Institutions of Public Housing Policies" allowed their *de jure* expansion into non-social fields. Since this change, the former public housing economy has followed the motto, "from state ownership to market orientation" (see e.g. Wolff (2003) for a historic overview of the development of German housing policy.). Liberalisation trends were continued during the 1990's, including drastic reduction of public funds for low-cost public housing and a new emphasis on private home ownership. Today public funding for housing is for all intents and purposes non-existent. These policies brought on an accelerated reduction of the number of rent and occupation regulated housing units available. Currently, such regulations still exist for approximately 2 million housing units. Although communities attempt to purchase rights to new regulations, the present financial situation inhibits any meaningful results. At the same time, rental regulations for individual projects are being prematurely absolved through early repayment of credit. According to surveys, in the next three years approximately 50% of the existing controls could be annulled and the construction of low-cost rental housing has been reduced significantly in recent years)

Today, housing developers are free from regulation in their business strategies concerning the content, in type and structure, of the use of their housing inventory. Construction can be financed through public or private funding, with only a stipulation of minimum use of private capital for public-financed housing.

Public low-cost housing is regulated by the Housing Subsidisation Law ("Gesetz über die Soziale Wohnraumförderung") as well as the Housing Use-Limitation Law ("Gesetz zur Sicherung der Zweckbestimmung von Sozialwohnungen") as last revised, 2001. Especially important is the "Housing Use-Limitation Law", which regulates rent controls, tenants rights, etc.

Rent setting

The difference between market-oriented rents and so-called cost-based and allowable rents should be defined. Cost-based rents for public housing are established according to an economic calculation. Through this inflexible process, which excludes market-price fluctuations, costs are covered for new public housing (costs for external and private capital, management, depreciation, administration, etc.) according to the Second Housing-Construction Law ("II. Wohnungsbaugesetz"). Calculations become inaccurate when costs, e.g. for maintenance, are fixed through legislation at an artificially low rate. If the cost-based rent is established by a local public administration, it becomes the allowable maximum rent. Both calculated rates do not necessarily coincide. If the allowable rent is lower than the cost-based rents, the investor is required to accept a so-called "effort-waiver" (*Aufwandsverzicht*). Due to its inflexibility, the concept of cost-based rents, also used as a guideline in privately financed housing construction, can lead to distortion of rents in the public housing sector. The concept also tends to cover fully not only fixed, but also variable costs, thus immediately eliminating competetive, price-regulating mechanisms.

Rent control responsibility is assumed by the individual German federal states. The establishment of rental limits is thus a political decision. In the past, costs of production, management and capital often reached sums of 100% more than the actually paid rents. The difference between real costs and paid rents is the resulting rate of subsidisation. The developer's share of private capital is low, and draws interest at about 4%.

When the minimum time-limit for rent regulation has ended (duration varies according to assistance program), the developer can raise the rents up to 10% in three years. After this period, the rents are established according to communal rental statistics or local comparable rates ("ortsüblichen Vergleichsmiete").

Allocations

No governmental regulations are available regarding housing construction for any certain social groups. Any business can use public funds, if it agrees to the regulations and limitations for the use of these funds: building standards, rent limitation, communal tenant stipulation. Regulations which influence allocation are gradually losing their effect, as public housing funding practically ceases to exist. Without state support private developers do not construct rental housing.

Sales

Principally, housing developers are allowed to sell their products at will to other private companies or to individuals. However, the following legal (depending on the legal form of the developer and the type of housing funding used in each case) and political limits must still be observed:

- Obligations, stipulated in the original funding contract, remain unchanged (occupation, rent), even after sale of the unit. A private developer must also observe these obligations. For example, if a private developer purchases a public housing unit for his own use, he is still obliged to observe the tenants' occupational rights guaranteed by law;
- If a housing co-operative enjoys tax advantages but sells apartments for a profit, it loses its fiscal privileges;
- Housing, which through military conversion becomes available to the state, can be sold at different prices, depending upon the extent to which the purchaser agrees to accept certain social obligations.

But, for housing, the administrative, communal-political and mental context is more important than the legal prerequisites. Politically, as housing policies have changed to one of ownership-support, purchase priority is given to tenants. But housing companies do not favor tenant purchases, because of the large effort involved, especially, for example, when all apartments in one project are not sold. Although it is a part of government policy and even subsidised in eastern Germany, sale to individual tenants is not a part of developers' current strategies. On the other hand, today, housing belonging to former quasi-public housing developers, especially in communities, is being sold on a large scale to national and foreign investment groups. These investors, with regard to long-range developments in the real estate market, are prepared to accept the required legal social obligations. Communal public housing is seldom sold on the market (for political reasons). In contrast, state owned housing is being extensively privatised.

Maintenance and renewal

Developers of public housing are required, as a part of the subsidisation program, to set aside defined funds for maintenance and renewal (Instandhaltungsrücklage), but without stipulations for the use of these funds. In addition, state and communities provide special programs for renewal, especially in the realms of environmental protection (e.g. heat insulation) and provisions for handicapped.

As of several years, national, federal-state and community governments, acting together, have developed an extensive program, called "The Social City" ("Die Soziale Stadt"), for the renewal of complete housing projects, in which developers receive substantial aid. In eastern Germany, and to a certain extent in some western German communities, financial assistance is being provided for the demolition of pre-fabricated tenements dating from the period of the German Democratic Republic.

8.2.2 Supervision

Control mechanisms differ according to the legal status of the different housing developers, whether limited corporation, tax-exempt landlord co-operative or tenant co-operative. An annual report and certain controls are legally established (e.g. § 53 *Genossenschaftsgesetz*) and certified through review boards.

In the state and communally owned development companies, the managing director presents an annual report for approval to the supervisory board, in which political representatives also participate. In co-operatives, the elected, honorary board controls the organisation's management and management's report is published annually (see figure 8.1).

Developers, such as co-operatives, which enjoy special tax status, are controlled by the finance authorities. When public funds for renewal or new construction are employed, control is performed by the appropriate ministries or their special agencies.

8.2.3 Financial Support

As already depicted, since the middle of the 1990's direct state financial aid to public housing has been drastically reduced. Governmental interests changed course, concentrating on subsidised ownership programs. At the same time a change from "object" subsidation to "subject" (housing funding or

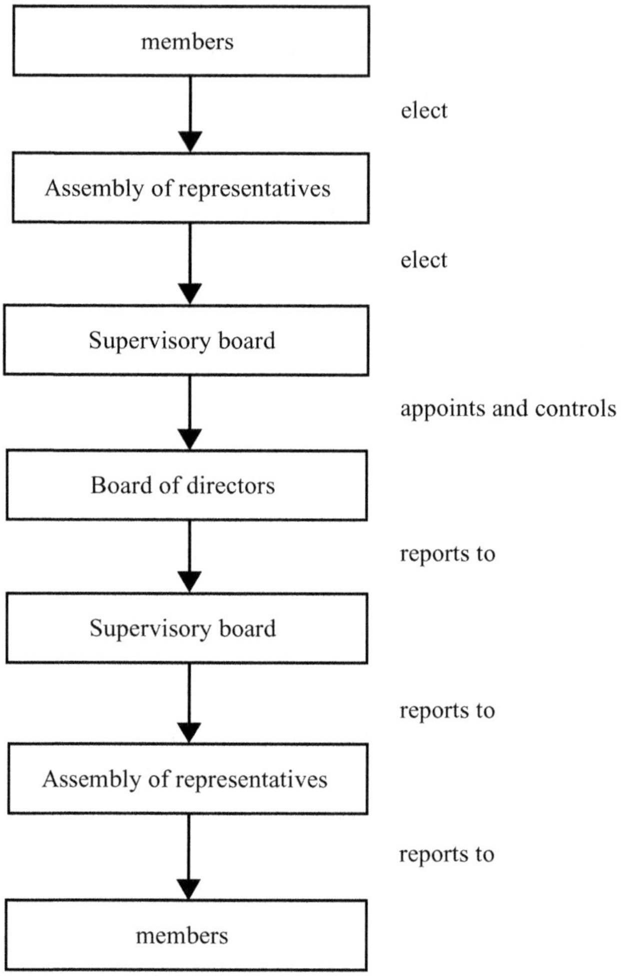

Figure 8.1 Example of checks and balances in a co-operative.

"Wohngeld") took place. Nevertheless, special state and communal assistance programs also exist, for example:

– ecological building methods (energy saving);
– construction of housing for handicapped;
– renewal or demolition of inferior pre-fabricated housing;
– purchase of regulations of housing for special social groups, etc.

Well administered housing developers are capable of making a profit and investing in new construction and renewal. However, depending on location, cost-rentals are now often higher than market rates. Therefore, particucarly in the case of new construciton, developers are waiting for 'better times', meaning a renaissance of public housing programs.

Reserves have been accumulated, through the use of funds allocated to maintenance. Today it is permissable to use these funds "horizontally", that is, for other housing units, than those from which the reserves come. New capital reserves are generated through the planning and sale of private homes and condominium apartments. In this case, the former quasi-public housing developers have certain advantages over private developers. The former often own originally cheap, but presently high valued real estate. Additionally, as former quasi-public developers, they still enjoy their clients' confidence.

8.2.4 Housing Market

On a national level, the German housing market is considered to be 'balanced'. The number of households and housing units are equal. Accepted is the fact that the population in Germany in the next 20 years will decrease by several millions, and that immigration will not compensate for this difference. Given these premises, there would be no need for new housing construction, even considering an annual loss of 100,000 units. In addition, momentarily, the demand on the part of families with children for apartments (flats), the dominant housing form offered by developers, is negligible. Rental single family dwellings are, for all intents and purposes, not a part of developers' portfolios.

On a regional level, the above statements have little meaning, since important locational differences exist. For example, in eastern Germany more than one million apartments are vacant, southern German urban agglomerations are faced with a housing shortage.

8.3 ASSET MANAGEMENT PRACTICE

8.3.1 Rent Setting

Since the repeal of the Public-Use Housing Law developers have the possibility of orienting their rental programs to market needs, further benefiting from legal margins for rental increases, when other rent controls are not involved. Up till now, this alternative has remained mostly theoretical, as in parallel with the newly won freedoms, the traditions of the individual developers were not eliminated. The new housing market orientation is not so much evidenced in developers' rental policies and is reflected mostly in the generation of new market fields and forms.

General statements concerning German developers' rental policies are hard too make. Differences between eastern and western, as well as northern and southern Germany are too large. In eastern Germany, as already stated, over

1 million housing units are vacant. Developers offer special bonuses to potential tenants, such as rental payments beginning in second month; free refrigerators, washing machines, etc.

Regarding policies under 'normal' market conditions, developers are, generally speaking, not always interested in full utilisation of the legal margins possible in unregulated housing. Principle guidelines are rental policies, directed towards long term rent regulation and binding the tenants, but differentiated according to tenant composition. Price evolution is controlled by housing companies so that levels remain under the ultimate achievable, keeping in mind increasing rental risks. On the other hand, these companies raise their rates to levels which assure dwellings' financing. As a rule, tenant changes lead to rental losses of three or more months duration due, partly, to substantial repairs, low demand, inflexibility of management. Results: high fluctuations which can leading to millions in losses for larger developers.

Contrasting with economical rationality, communally owned development companies generally keep a segment of low-cost housing units in reserve, as this is expected or required by the community. The older co-operatives limit rent-increases out of tradition, since higher rents would be unacceptable for their tenants, often long-time members. Both of the above policies lead to a reduction in standards beneath the level of a younger generation's expectations (larger and better equipped units). Developers, in these cases, are caught in a dilemna. Young co-operatives, having achieved a certain inventory of housing units, attempt to keep the initial rents under the approved levels, through excellence in planning and management and to limit auxiliary costs through self-administration.

8.3.2 Allocation

According to the Housing Regulation Law ("Wohnungsbindungsgesetz"), low-cost housing is subject to occupation limitations. Depending on the type of subsidisation, maturity of contract, individual Federal state laws and regulations, communities can independently determine housing occupation, or the landlord can decide from a list of candidates. The community issues the housing occupation license. Additional needs-criteria are designed to assure that, for example, certain socially disadvantaged groups, such as single-parent families, receive adequate housing.

The unregulated housing belonging to the former quasi-public housing companies is made available on the open market. For regulated apartments, community housing agencies possess allocation rights. These departments maintain lists, categorised according to needs, of housing applicants, who, in turn, receive a so-called "Housing Rights License" ("Wohnberechtigungsschein"). Families, faced with eviction and possible homelessness, as well as senior citizens, needing housing for handicapped, enjoy highest priority.

Detailed regulation varies between communities. The public agencies submit three tenant candidates to the developer. Communication between developer and agency takes place on an informal basis. Hence, the agency representative knows exactly which tenant is to be suggested for a specific apartment. The communal agency's waiting list is often shorter than apparent, as candidates often remain registered, independent of their future status.

Tenants may retain their favorable rent conditions, even when their income level considerably increases beyond the limits allowed for the occupation of public housing. Strict adherence to regulation could endanger complete housing occupation.

Since an increasing number of regulating contracts are expiring, communities attempt, through modernisation assistance, to renew regulations. Whereas in Germany, 1989, approximately 3 million apartments had restricted occupation, by 2000, only 1,5 million were still regulated.

Some co-operatives, oriented towards exclusiveness, refused to take advantage of state funding, not subordinating themselves to the implicit regulations. But, in fact, occupational rights are informally negotiated. The developers cultivate relations with the tenant-registering agencies. Agency employees know which tenants are best suited for developer and project. This informal process is the initial step in pre-selection. Primary criterion: occupation should not create social conflict, neither in the housing project, nor in the apartment building. Through these methods, 'good' neighbourhoods remain 'clean', whereas disadvantaged tenants are relegated to other areas. For the latter, special programs, such as the "Social City", are developed, through which the personal costs for revitalisation are funded.

In the case of co-operatives, depending on their type and size, a more informal housing market exists, independent of their housings' regulation status. The co-operatives justify this situation with their inherent communal obligation, with a certain exclusiveness as prerequisite.

8.3.3 Sales

Until recently, the sale of rental units to private investors (mentioned here as 'privatisation') was considered a break with the traditions of the quasi-public housing developers. But within a few years a transformation has taken place. Some even claim that developers compete in the realm of privatisation. The reasons for privatisation differ amongst developers.

- Developers associated with industry have always been attempting to mobilise their frozen assets. Now they are accomplishing this feat, working as real estate agents, not only selling larger agglomerates, but also buying, re-subdividing and re-selling complete complexes. Some call these methods 'transit trade';

- Community-owned developers are required to give up a part of their housing, thereby strengthening the communal treasury;
- Other developers sell their inventories, thus restructuring their remaining portfolios for more rational administration.

Developers avoid direct sales to tenants, as this involves excessive effort, and thus trade directly with intermediaries, who then assume the difficulties and inconveniences of these transactions.

According to InVIS (2003) the the real estate market currently focuses on the approximately 5.85 million apartments, around 20% of the total, owned by commercial landlords. Only in this sector, there is an interest in capital voltility, whereas smaller and medium-sized developers and co-operatives are more concerned about stability, considering housing sales to be a violation of their social obligation.

Municipalities regard the sale to tenants positively. According to their thoughts, housing purchases by affluent clients lead to social stabilisation in otherwise from deterioration threatened neighborhoods, but results are often disappointing since many tenants will not or cannot assume ownership responsibilities.

In addition to regular sales, alternative forms of ownership are recurringly discussed (rental-purchase, formation of mini-co-operatives, etc.), but such models have achieved little quantitative relevance.

8.3.4 Maintenance and Renewal

Generally speaking, the efforts of German housing economy management have resulted in a relative continual dwelling-quality maintenance, which in turn guarantees both rental income and a socially acceptable living standard. However, this quality exists only in the western part of Germany and even in western Germany a certain stagnation in maintenance has occurred. Increasingly, due to inferior standards, apartments remain vacant. Landlords who have neglected continual maintenance and fail to heed needs and demand, must accept present and future vacancies, even to the point of endangering their capital investment.

Many renovation projects are conducted with little regard for the tenants' social strata. Despite this fact, housing for elderly tenants is less liable to renovation than those of younger inhabitants.

Improvement by tenants are regarded as undesirable, as extensive organisation is required. In examples where tenants have constructed their own bathrooms, conflicts have arisen. Even tenant participation in maintenance is not allowed, since it is supposed to be too complicated.

Technical deficiencies are the principle causes of renovation and maintenance. In the process of maintenance, generally, a higher quality is obtained of bathrooms, kitchens, stairwells, heat insulation, windows, new balconies, etc. Thus, not only the substance is preserved, but also the market value increased. However, when we critically look at examples in publications of the housing industry we note that landlords' maintenance strategies seldom attempt improvements that might attract new groups on the market. The private investor will only attempt more fundamental improvements in centrally located urban projects, obtained from public housing developers. Luxurious modernisation, with negative effects on existing tenants, is not practiced by public-housing developers. Nevertheless, more radical strategies regarding existing housing are being developed. As new housing construction has decreased, city planning experiences a paradigm change, under the motto, "The future lies in the existing structure". The housing industry took up the banner, and, even in western Germany, after years of orientation on new construction, the emphasis was transformed to maintenance and renewal of existing buildings.

Maintenance is a substantial problem for landlords with an inventory of high-rise buildings from the 1960s–1970s, in which reconstruction of elevators, façades and roofs are required, without a change to higher-income tenants. The renovation of now century-old buildings, dating from the beginnings of the co-operative period, also presents a huge financial challenge.

Example NH Frankfurt: Manual and Schooling

Chosen from the wide spectrum of existing German entrepreneural forms, we discuss the business practice of a large company with 43,700 housing units and headquarters in Frankfurt/Main: the Nassauische Heimstätte GmbH (NH). Within the field of housing administration, NH Frankfurt is probably the most innovative housing company in Germany. Centralised and decentralised administrations are systematically interconnected. The NH has developed a manual for each housing administrative procedure. Every step proceeds according to an established pattern. When, for example, a tenant gives notice, the chronological order of the resulting administrative procedure follows a defined pattern, from formalities to actual dwelling renovation. Central management can participate in every phase of this process. In hundreds of pages the manual describes and depicts, through graphic descriptions, each administrative measure.

As the many of its tenants are older and immigrant, the NH has developed its own schooling program for maintenance workers involved with renovation. These workers learn to act with social competence, according to certain behavioral rules, for instance: to arrange appointments in advance, per telephone; to leave an apartment cleaned upon work completion; to take consideration for the aged; etc.

8.3.5 Decision-Making Process: Considerations, Methods and Strategies

Since the revoking of the Public-Use Housing Law, housing developers' business policies have undergone radical transformation. Combined with the current structural changes, this process has become even more dynamic. On the one hand, housing companies are required to act freely according to market demands. On the other hand, socially oriented politicians request a continuation of the tradition of public housing provision. These developers are challenged to fulfill higher social standards in their 'products' and management, than those of private companies. Such housing companies are informally and formally limited by political objectives. Enterprises which have worked over decades in partnership with state and communal housing agencies are expected to provide target social groups with 'appropriate' housing. These developers, along with the social service companies, are still partners in social policies.

Many developers reacted long ago to the new situation, for example, through new 'products', more differentiated than the original core housing; through diversification into other branches of the business; through exposure in fields of urban development. Some of these companies are even attempting to export their know-how to other European countries, especially to the new eastern European countries.

Locked into traditions

The current social structure transformation accentuates the weaknesses of those housing companies which molded their strategies according to the requirements of governmental market intervention and communal policies. Even today, housing companies wait for state intervention through the program, "The Social City", before commencing revitalisation involving significant social conflict in lower market segment housing projects. In industry-oriented companies, modern management forms were only introduced after owners formulated profit-oriented requirements. These activities often resulted in the sale of apartments.

Numerous expertises, little effect

Although often without noticeable strategy, investigations and expertises endlessly proliferate, contracted out by companies and their professional associations, seeking support for investment decisions. Such expertises are produced by, for example, "InWis", with its connection to professional associations, or *Analysen und Konzepte* ("Analyses and Concepts"), close to co-operatives. In any case either private or university institutes are involved. To what extent the results of these manifold expertises have significant influence on strategy, can hardly be seen, even in specific cases. The insignificance of such analyses becomes more evident, in the fact that expertises often are first published not before, but after establishment of the affected decision making process.

Decisions on the basis of personal opinions

At the level of housing developer leadership many choices are not based on rational, economic knowledge, but rather founded in the decision maker's mentality or political orientation. In the case of community developers, political direction often dominates over economic criteria. Older-generation managers, express concerns, that the new generation of managers 'unfortunately' have separated themselves from the social principles, which enabled the company's success, only to replace them with 'neo-liberal' ideas.

Dependency upon social-democratic policies or christian ethics

Middle-sized and larger former quasi-public housing companies, once owned by communal or government entities, even some co-operatives, seem to understand themselves as exclaves of the executive, as evidenced in annual reports, congress speeches and personal experience. On the municipal level, the housing corporations are often the last refuge of creative political action, after the privatisation of essentially all public utilities. The housing company directors recruit executives from SPD-circles. Employees consider themselves quasi-civil servants. Shop stewards can use their influence in real terms. High employment in the development companies was considered positive; economic efficiency less so. The enterprise's administration buildings are hardly distinguishable from public edifices. Only in recent years has modernisation taken place, through new structuring of departments and resulting transparent and decentralised organisation.

Middle-sized and larger co-operatives reveal similar bonds to the social-democratic tradition. Although directors and management actually make all decisions, principles of democracy and self-administration are followed. The numerous religious housing companies also practice intensively the social precepts of their respective churches.

Summarised, the strategy of housing developers, in which state and communities still participate, even that of larger traditional co-operatives without public shareholding, is dominated by political interests, despite an increasing economic emphasis. This development does not mean that, politically, direct influence is taken on the business process. Rather, political interests are quasi internalised. Despite this fact, many enterprises are capable of conducting their administration effectively and rationally.

Structuring the decision making process

As a consequence of housing companies' liberation from the shackles of the quasi-public economy and governmental regimentation, further development through procedurale models for modern management would seem a logical following. Such prototypes are being developed in universities which have incorporated housing economy in their curricula. Recently, special departments for real estate economics have been introduced. However, an investigation of the housing economy's central publication, "Die Wohnungswirtschaft", reveals how little such models are actually used in practice. One reason for this deficiency lies in the history and tradition of these enterprises, which have yet to accustom themselves to their independence from governmental housing policies. Legal regulations (Housing Use Limitation Law or Wohnungsbindungsgesetz) once defined precisely the developers' commercial freedom. Decision-making processes (such as those involving the annual production of housing units, investments in maintenance, etc.) depended upon state assistance policies, and, in the case of communal developers, from local housing and social policies.

A similar statement can be made about the use of market analyses. In Germany, a large number of private and public institutes exist, continually publishing analyses, forecasts and concepts, not only in overall, but also in detailed situations (e.g. regional market studies). Even middle-sized and larger cities contracted out numerous studies of the communal housing market up until a few years ago. However, many of these documents were never used, as the dynamics of real development overtook the slower pace of analyses.

Cooperation with Experts and Scientists in Asset Management

NH Frankfurt am Main belongs definitely to the small group of housing companies which enjoys the "luxury" of co-operation with external experts and with the local university. For example, an advisory board composed of representatives from political and scientific bodies, has not only representative functions, but also gives stimulus to the enterprise.

The developer awards two prizes for outstanding student projects and scientific studies (Ernst May-Preis, Landmann Preis), enabling the company to maintain contacts with the scientific community.

Portfolio optimisation versus social obligation

Today, the question of the ideal company portfolio composition is a subject of utmost importance for housing developers and co-operatives. Here, conflicts arise between market and social obligation, a concept which has been continually less rigidly defined.

Most housing companies' holdings are composed of a jumbled mixture containing different apartments of different vintages, including low-rise buildings dating from end of 19th century, to end of the 1970's high-rise. In addition, these buildings can be variously located, either concentrated within urban boundaries or distributed throughout a whole region. Many holdings are suffering from dramatic deterioration. Many portfolios exhibit heterogeneous legal status. Situations exist, in which neighbouring buildings rent at different rates, only because their construction phases took place in periods of different housing assistance programs.

Strategic planning requires an evaluation of the company's substance and an estimation of the potential of it's housing stock, but practice seems to be mostly developed within the consultancy sector, instead of within the housing developers. Numerous consulting firms advertise their services to the housing economy through typical slogans, such as "intelligent business solutions", "process optimisation", "risk management" and "portfolio analyses" focus interest, and, reinforced with appropriate software, prototypes are being developed, even in scientific fields. The services offered by real estate administrators become more and more complex and complicated. Such firms offer the client with large housing inventories new services beyond pure analyses, such as portfolio management and strategy consulting. The terminology stems from asset management vocabularies. Emphasis lies in the premise: a well diversified portfolio reduces the risk posed by individual objects. (See e.g. Wolfgang Pelzl, 2002)

In contrast to the consultancy practice, many housing developers are neophytes in the use of portfolio analyses and even more so in the realm of portfolio management. The companies know exactly the number of housing units in their possession, their floor area, their state of repair, but the future value of the housing units is judged by management's individual, personal estimates. This is can be explained partly because portfolio models are mainly designed for large real estate investment companies and not for the traditional housing economy. For the latter companies these models are too technocratic. Furthermore, portfolio management can only be commenced when an all-inclusive, long-term real property data bank has been established, a very expensive process. Thus, most likely, the decision-making process in the public housing sector will remain unchanged, whereas the administrative practice will continue to be streamlined.

8.4 CONCLUSION

The spectrum of housing developers in German public housing shows a lot of variety. The bulk of the firms are made up of former quasi-public companies, now in communal, state or private ownership. Fewer are the co-operatives, with their differing structures. Lastly, the private housing economy also operates in the field of public housing. The housing companies and co-operatives own over six million of the approximately 32 million apartment units in Germany.

Today, the housing economy stands on the threshold of significant change. Although the developers, as a result of the revoking of the Public Housing-Use Law, conduct business *de jure* according to market principles, only until the most recent phase of structural transformation, has the meaning of this paradigm change begun to function, and have its effect on management. In many enterprises, especially communal companies, the motto, "from government agency to the market" has not been completely put into practice. The results are to be seen in developers' administration structures and management, as well as the orientation to communal and state housing policies. The latter policies, however, have almost completely departed from public housing assistance programs. Now the task at hand is: preservation of existing housing.

Business policies hover between the contradictory poles of the political requirement of observing market demands, on the one hand (e.g. communal companies should be profitable), and the fulfilment of social requirements, on the other. As a result, housing developers are expected to provide housing in the lower market brackets. At the same time, for more and more existing apartments, the occupation regulations and rent limits, stipulated by the laws for public housing assistance, are expiring. But housing companies decline to exploit the legally allowable rent increases for unrestricted rentals, as, first, they cannot afford further pressure on the lower market segment, and, second, continuity in these socially stabile housing complexes takes priority over high profit.

Just as any business firm, theoretically, is allowed to construct public housing through the use of public funding, they can also sell them. All accompanying non-expired regulations remain valid, and must continue to be accepted. However, most housing companies and co-operatives regard sale of rental units as a taboo, although state and communal governments, because of the public debt emergency, question this direction. Thus, in recent times, large housing properties belonging to the German Railroad, postal services and other former public institutions have been offered up for sale. In addition, attempts to restructure portfolios could lead to further sales of rental units. On the other hand, because of certain complications, tenant purchases are minimal.

Housing companies possess considerable inventories, which, in the case of co-operatives, result from 100 years' accumulation. The resulting architectural, even legal heterogeneity and dispersion impede a rational administrative management. In addition, decades of concentration on new housing construction resulted in neglect of inventory reorganisation. As housing companies and their federations have difficulties with distancing themselves from governmental intervention, and have come to expect improvement in their situation through recurring smaller housing crises, they have neglected market developments. Housing economy leaders thought, through installation of bathrooms and central heating, application of exterior heat insulation, etcetera, the dwelling stock could be satisfactorily improved to future norms, but this policy has proved to be partly deceptive. For young, ambitious families, life in public housing, at housing company levels, fails to meet their living standards. Concerning housing projects in the lower market sectors, the mentality of the former quasi-public housing companies is evidenced in the fact that only subsequent to the requisites of the governmental program, "The Social City" have they become active in social management. Even housing co-operatives with one hundred years of tradition are forced to accept housing vacancies because their apartments and housing projects no longer meet current standards—and not only building standards! Many developers reacted long ago to this evolving situation, for example through economic expansion from their original core operations to more differentiated, housing-associated activities; through diversification in other fields; through prominence in areas of urban development. However, most companies are locked in the chains of immobile tradition.

A wealth in studies and analyses relevant to asset management exists, contracted by housing associations to associated institutions. Terms such as "portfolio management", "client orientation", "process management", have found their niche in the housing economy vocabulary. Nevertheless, regarding developers' policy determining processes, it becomes evident that decision-making stems from head management's personal opinions, and (in communal or in state-owned companies) from the owner's political direction. Accepting management activity reports in the central publication, "Die Wohnungswirtschaft") of the Housing Economy Association (Gesamtverband der Wohnungswirtschaft) as an indicator, it would appear that housing companies with an effective and socially oriented management, such as the Nassauische Heimstätte in Frankfurt/Main belong to a minority. The bridging of the gap between social responsibility and the challenge of the market economy will still require much time.

REFERENCES

Bone-Winkel, S. (1994) Das strategische Management von offenen Immobilienfonds unter besonderer Berücksichtigung der Projektentwicklung von Gewerbeimmobilien, in: Schulte, K.W. (ed.), Schriften zur Immobilienökonomie, Vol. 1, Köln, Immobilien-Informationsverlag Rudolf Müller.

Bone-Winkel, S. (2000) Immobilienportfolio-Management. In: Schulte, K.W. et al. (eds.), Immobilienökonomie, Vol. 1: Betriebswirtschaftliche Grundlagen (2nd edition), München/Vienna, R. Oldenburg-Verlag, pp. 765–811.

Bundesamt für Bauwesen und Raumordnung (2001) Wohnungsprognose 2015, Bonn.

Kühne-Büning, Lidwina, u.a. (1999) Zwischen Angebot und Nachfrage, zwischen Regulierung und Konjunktur. Die Entwicklung der Wohnungsmärkte in der Bundesrepublik 1949–1989/ 1990–1998, in: Wüstenrot-Stiftung (ed.), Geschichte des Wohnens, Vol. 5, Stuttgart, Deutsche Verlags-Anstalt, pp. 153–232.

Pelzl, W. and Wellner, K. (2003) Emtwicklung eines Portfolio-Management-Management-Systems, Norderstedt, Verlag BoD-GmbH.

von Beyme, K. (1999) Wohnen und Politik, in: Wüstenrot-Stiftung (Hrsg.) Geschichte des Wohnens, Vol. 5, Stuttgart, Deutsche Verlags-Anstalt, pp. 81–152.

Wolff, A. (2003) Eine kleine Geschichte der deutschen Wohnungspolitik seit dem Zweiten Weltkrieg, vhW forum wohneigentum, 2 April 2003.

9 LATVIA[1]

SASHA TSENKOVA

9.1 SOCIAL LANDLORDS IN LATVIA

Latvia has the legacy of an excessively controlled housing system, which operated from 1940 to 1991. The system was based on low housing costs, centralized production and state control over housing allocation. Home ownership was discouraged, all housing was social and universally affordable due to extensive subsidies and macroeconomic regulation of prices. Large volumes of social housing were produced, while little was invested in the maintenance and upkeep of the existing stock (Tsenkova, 2000a).

The 1990s have marked a major departure from the Soviet model with deregulation of housing markets and privatisation of social housing being the flagship of the reform process. In the context of this wholesale shift away from direct state intervention to market-based provision of housing services, municipalities have emerged as 'the new social landlords' with major responsibilities for the provision of social housing as well as housing assistance (Tsenkova 2000c). The organization of social housing in Latvia still exhibits the features of a 'command system' where ownership and management is vested with municipalities and pricing policies are not sensitive to demand or quality of housing services (Kemeny 1995). Allocation decisions rely on bureaucratic processes and institutional arrangements (relating to allocation and rents) seem to impede the transition to a more market sensitive model.

The most recent data on tenure composition in Latvia reveal some interesting patterns. Table 9.1 compares the ownership of residential buildings and

[1] **Acknowledgement**

The material for this chapter is based on research carried out in the framework of the Latvian-Canadian Housing Cooperation Project. The financial support of the Canadian International Development Agency (CIDA) and the University of Calgary is gratefully acknowledged. Opinions expressed in this report are those of the author and cannot be attributed to CIDA. During my field work, I had insightful discussions with housing policy experts, municipal officials and leaders of different housing institutions. I thank them all for their cooperation and professional assistance.

Vincent Gruis and Nico Nieboer (eds.),
Asset Management in the Social Rented Sector, 161–184.

Table 9.1 Ownership of housing in Latvia (2000)

Form of ownership	Residential Buildings[1] %	Individual dwellings[2] %
Public sector:	93.0	30.0
State	17.0	0.5
Municipal	76.0	29.5
Private sector	7.0	70.0
Owners	Not applicable	60.0
Housing cooperatives/ Home owners' associations	2.5	1.0
Other[3]	4.5	8.5
Total	100	100

Sources: author's estimates.
Notes:
[1]Based on data in Central Statistical Bureau, 2001a.
[2]Based on data in Central Statistical Bureau, 2002a.
[3]The category 'other' includes institutional investors as well as private landlords.

individual dwelling units. Public ownership (by state enterprises, municipalities and state institutions) is as high as 93% and municipalities have the lion's share of the housing stock (76%). Just 7% of all residential buildings are owned by private institutions such as housing cooperatives. However, private ownership of individual units has reached 60% in 2000 and it might be even higher if we consider the share of housing cooperatives (1%) and the growing importance of the privately rented housing included in the category 'other'. Publicly owned housing is 30% and continues to decline. What this means is that private units (rental or owner occupied) are typically found in publicly owned buildings. While the transfer of property rights related to the building envelope and the land in these cases might take years to complete and might be considered 'a transitional element', the situation nevertheless poses some very unique problems for asset management in urban areas where most of the multi-storey housing in quasi-public form of ownership tends to be located. In the aftermath of mass privatisation of social housing, which is discussed in more detail later, social landlords in Latvia still have a much broader mandate and some unique challenges to cope with.

At the moment, the residential properties owned by the public sector are the home of a wide spectrum of the population. In fact, due to former communist policies of housing allocation, in most of the cases one could find a social mix in all neighbourhoods and buildings. The new definition of social housing requires vacant units to be allocated to socially disadvantaged and low-income households.[2] This appears to be one of the most challenging tasks for Latvian municipalities in the future. Previously highly dependent on central government control, the 'new social landlords' have become locally regulated and accountable to municipal councils. The institutional reforms in the housing system and the new financial regime for operation allow more autonomy in decision-making but also imply a growing social responsibility to deal with poverty and social stress. These legally binding rules also imply that social landlords are not expected to build owner-occupied housing or invest in rental housing for higher income groups. On the other hand, as mentioned previously, rent structures are not sensitive to demand and there is no mechanism for exit from the sector if the household's income increases above a certain threshold.

9.2 NATIONAL CONTEXT FOR ASSET MANAGEMENT

9.2.1 Regulations

The following sections provide a brief description of the national context with a focus on regulations related to rents, allocation, and privatisation of municipal social housing. The impact of these regulations will be explored in a comparative perspective drawing on information related to the large urban centres in Latvia.

Rent setting

Municipalities are entitled to set rents in their administrative boundaries, which is a radical departure from the previous policy of state controlled rent levels. The legislation also allows rents to be differentiated according to quality, location of housing and tenant's income.

The data on municipal rents in the seven republican cities indicate significant differentiation with rents in Venspils and Jurmala 50% higher than the average (Table 9.2). In practice, Riga and most municipalities in Latvia are charging uniform rents within their jurisdiction, thus pooling costs across the stock. With the exception of one municipality—Jelgava—the quality of housing is not a factor determining municipal rent levels. Rents barely cover operation costs. Correspondingly, maintenance companies carry out marginal upkeep and resort to patchwork maintenance and emergency repairs.

---

[2] This would include units that have not been privatized by the tenants, or any other legal entity, which become vacant as a result of turnover. The other additions to the social housing portfolio come from conversion of non-residential properties and/or acquisition of new social housing.

Table 9.2 Rent differentiation in municipalities (2002)[3]

Municipality	Lats per sq.m
Daugavpils	0.082
Jelgava	0.12–0.17
Jūrmala	0.15
Liepāja	0.11
Rēzekne	0.12
Riga	0.14
Ventspils	0.18

Source: Central Housing Privatisation Commission Survey, June 2002.

Allocations

The Law on State and Municipal Assistance for Resolving Housing Problems, adopted in 1993, specifies a set of eligibility criteria for the registration and allocation of social housing. Households are categorised in three groups:

- large families with disabled children, or large families where parents have retired; and politically repressed individuals;
- households in denationalised properties where one flat has to be vacated for the needs of the owner;
- households in communal flats or households living for 40 years in units without adequate utilities, families sharing one flat, households in substandard units.

Priority is given to households with special needs: orphans, the handicapped, chronically ill, the elderly and single parents.

Social housing in Latvia can be defined as the property of municipalities that is exempt from privatisation and can be rented to low income, disadvantaged households in accordance with the Law on Social Assistance. In 1999, there were 461 social apartments in Latvia with 3,026 inhabitants (MoEPRD, 2001). Although Riga's share is disproportionately high compared to the rest of the country, social housing units (431 in total) constitute less than 0.25% of the city's total housing stock (Social Development Fund, 2001).

Sales: the 'right to buy'

The Law on the Privatisation of State and Municipal Apartment Buildings, adopted in June 1995, regulates privatisation of state and municipal housing.

[3]Note: The local currency is pegged to the EUR at a rate of 0.648 Lat = € 1.

Two distinct privatisation methods can be applied: i) privatisation of the entire residential building with sale of individual units to existing tenants, which is followed by registration in the Land Book, and ii) accelerated privatisation, which allows privatisation of a housing unit within a multifamily building prior to privatisation of the entire building (Kursis, 1999).

Tenants in state and municipal housing can use privatisation vouchers—*privatizacijas sertifikati*—to purchase their unit. The value of a voucher officially was estimated at 28 Lats in the early 1990s, corresponding to 0.5 sq.m. of residential space. Latvian citizens were entitled to a certain number of vouchers depending on their age.[4] Assuming that certificates are bought at their current market value of 1.5 Lats, a 50 sq.m., a flat might cost 175 Lats, which is roughly 3.5 per cent of the market value of the average dwelling in Riga's real estate market. Vouchers can be used to privatise housing till January 1, 2004; this deadline has been extended several times in the last two years. It is politically very unpopular to impose a moratorium on housing privatization, although some other East European countries have done so. It is unclear how the privatization will proceed in the future and what course of action might be appropriate. Although privatisation in Latvia had a 'late start' compared to other transition economies, as Figure 9.1 demonstrates, it has led to an increase of 30% in private ownership of housing. Interviews with officials suggest that the privatization process has gained momentum and continues to reduce the housing portfolio of public landlords in Riga and other large urban centres.

The massive sale of public housing at below market prices is motivated by political considerations and economic efforts to restructure the housing sector on market principles. The policy designed by the central government on one hand is depriving municipalities of their valuable asset, and on the other, is creating a culture of home ownership that will take years to become efficient in housing management.

Maintenance and renewal

In accordance with the new legislation, the responsibility for housing maintenance and management is transferred to the home owners. Provided that 50% or more of the dwellings in a building are privatised, the owners decide on the

[4] Introduced in November 1992, the distribution system allocated one voucher per person for each year, plus another 15 vouchers for citizenship. The typical family of three (2 adults aged 40 years and a 15 year old child) could acquire 140 vouchers sufficient to buy a 70 sq.m. flat. The fact that privatisation vouchers could be used in the industrial privatisation, as well as in the purchase of pension funds, has lead to a market for vouchers where trading fetches historically low prices—between 1.5 to 2.8 Lats in 1999 (Ministry of Economy, 1999).

Table 9.3 Residential buildings transferred to owners for maintenance and management

	State owned buildings	Municipal buildings	Total
Multi-apartment buildings offered for privatisation	2,350	16,983	19,333
Including buildings where more than 50% of apartments have been privatised	1,500	10,428	11,928
% of total	64	61	62
Including buildings transferred to the owners for maintenance	724	814	1,538
% of total	31	4	7

Source: Central Dwelling Houses Privatisation Commission, 2002.

form of asset management. The Law on Apartment Ownership provides two alternatives: i) establishing of a home owner association; and ii) establishing of Agreement to contract a private or municipal maintenance company and/or carry maintenance on a self-help basis.

Table 9.3 summarises progress in the transfer of multi-apartment buildings for maintenance to home owners as of April 2002. While the possibility to complete the privatisation process exists in 62% of the buildings, only 7% have registered a home owners association or established an Agreement.

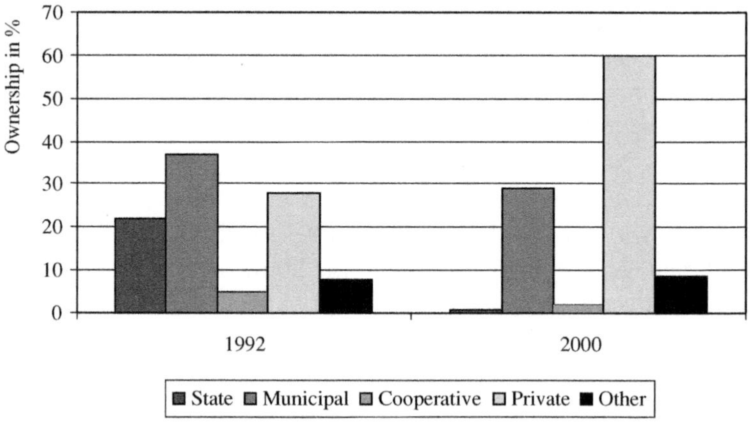

Figure 9.1 Housing privatisation in Latvia
Source: MoEPRD, 2001; Central Statistical Bureau, 2002a.
Note: The category 'other' for 1992 includes housing owned by institutions, while in 2000 'other' includes private rental housing.

This process has been much more successful in state owned housing. Interviews indicate considerable difficulties in large municipalities with housing concentrated in peripheral housing estates, where 'new owners' are reluctant to assume responsibility for common areas in the prefabricated buildings. The situation places municipal maintenance companies in the unique position of managing privately owned housing, while collecting uniform locally regulated rents for their services.

The legacy of Soviet housing policies has contributed to a serious backlog in maintenance and repair of the existing stock. The situation was further aggravated during the transition period due to government withdrawal from the financing of repair and modernization activities. According to a recent study by Latvian Ministry of Environmental Protection and Regional Development the prefabricated panel buildings, which comprise close to 59% of the housing stock in urban areas, require 450–490 million Lats for energy improvement and reconstruction (MoEPRD, 2001; Ministry of Finance, 2001). While surveys show that the economic effect of such improvement measures can be as high as 8–8.5 million Lats, the existing legal, institutional and financial barriers have led to further deterioration of the stock (COWI, 2002).

9.2.2 Supervision

At the central level, supervision of social landlords is limited to the development of general policy guidelines pertinent to social housing management. The Ministry of Local Government and Regional Development of Latvia prepares legislative documents regulating technical requirements, maintenance and rents. The Central Dwelling Houses Privatisation Committee oversees housing privatisation of state housing, carries supervision and control of the operation of the municipal privatisation committees and participates in the organisation of housing management after privatisation. The Ministry of Finance develops programs for housing investment and supervises the State Property Agency managing the state housing stock. The Ministry of Welfare establishes the social policy and develops legislative bills related to the provision of social housing assistance.

At the municipal level, social landlords are supervised by municipal councils.In general, Latvian municipalities have the following housing functions: i) maintain the municipal housing stock; ii) perform privatisation of the municipal housing stock; iii) house low income residents; and iv) provide social housing assistance. Most large municipalities have a very hierarchical structure where separate municipal departments carry out each of the four main functions. In the City of Riga, for example, the Housing Administration Division: i) maintains a housing waiting list; ii) rents vacant flats; prepares flats for auction. The Communal Economy Administration: i) prepares operating and capital budgets for review by Riga City Council and implement such budgets through management and maintenance activities; and ii) collects utility and rent payments for residential and commercial properties. The District Executives supervise 21 housing maintenance organisations. All administrative units are under the direct supervision of a body of elected council members.

9.2.3 Financial Support

The Law on Social Assistance, adopted in 1995, eliminated state financial support for housing, delegated these responsibilities to municipalities and allowed them to determine locally the income level for housing benefits. This emphasis on local autonomy, essentially beneficial for the development of democratic, participatory governance, has left a lot of unfunded mandates. Small municipalities have difficulties carrying out essential services and certainly refrain from investment in social housing. The move to demand-based housing assistance is essentially driven by what landlords can afford, as opposed to rational responses to growing waiting lists for housing. Under the present regime of fiscal austerity, financial support is available to low income households regardless of their tenure status. However, the transition to markets, where households are paying the full cost of housing services, is manifested in significant affordability problems. The data indicate that 55% of all households, and 66% of the tenants, find it difficult to maintain their utility and maintenance payments.[5]

Reportedly demand-based housing assistance accounts for 35% of social spending in large municipalities. The data on expenditure for housing allowances in the seven republican cities show a modest decline over time (Liepaja and Jurmala are notable exceptions), mostly due to budget cuts. A survey administered to 15 municipalities in Latvia revealed a general consensus that: i) housing assistance is grossly inadequate compared to need, ii) criteria for low income cut-off are set at a very low level so that only the poorest of the poor can qualify, and iii) the funding, which usually amounts to 2 per cent of the municipal budget is limited and distributed 'on a first-come-first-serve basis', which does not mean that it reaches out to all households in need (Tsenkova, 2002). The average amount per person depends on the

[5] Without the extensive state support utility payments, particularly for heating, have become a major problem for owners and renters alike. Close to 10% of the households have previous debt with respect to rent/maintenance payments and in the rental sector this share is as high as 19.7% (Central Statistical Bureau of Latvia, 2002b). Rent arrears have become a wide spread phenomenon creating a lot of pressure on the administration and management of social rental housing.

Table 9.4 Distributional inefficiency and overcrowding

Number of residents in dwelling	Of which by number of rooms						
	1	2	3	4	5	6 or more	Total
1 person	82,589	77,508	27,055	6,687	1,997	998	197,367
2 person	39,055	114,385	51,077	12,047	3,780	1,540	222,277
3 person	15,000	77,245	55,088	13,294	4,290	1,697	166,969
4 person	6,113	44,785	50,376	15,719	5,398	2,214	124,854
5 person	1,698	13,081	20,730	9,730	4,052	1,826	51,229
6 person	492	3,874	7,256	4,523	2,166	1,258	19,610
7 person	327	2,220	4,234	3,215	1,917	1,432	13,394
Total	145,274	333,098	215,816	65,215	23,600	10,965	795,700

Source: Central Statistical Bureau, Census Data, 2002a.

financial position of individual municipalities and varies considerably across Latvia from a high of 39 Lats in Rezekne to a low of 20 Lats in Daugavpils (Social Development Fund, 2001).

Given the lack of supply-based financial support for asset management and inadequate assistance on the demand side, very few social landlords have borrowed for renovation and improvement of social housing. Under these conditions the financial position of social landlords no doubt improves with the stock-transfer through privatisation. The initiatives to 'match' social tasks with financial means are not surprisingly very limited and driven by municipal politics.

9.2.4 Housing Market

The concern with housing market performance in the first instance is with the relationship of the total number of housing units in the country to the number of households wishing to occupy separate units. Two indicators can measure housing consumption at the national and local level: number of units per 1,000 inhabitants and average space per person. Latvia has 375 dwellings per 1,000 inhabitants, which is close to the EU average, but the average space per person is as low as 21.8 m^2.

The next question relates to the appropriateness of housing supply with respect to family characteristics. The census data indicate that some households in Latvia are overcrowded, while others consume larger quantities of housing (refer to Table 9.4). To a large extent this is a legacy of the Soviet system of housing allocation, where the ratio of one person per room is the ideal target for distributional efficiency. But even this crude indicator for housing distribution reveals record high levels of overcrowding. Close to 65% of the people live in smaller units than the norm; the problem is particularly visible in the situations of one- and two-room dwellings. Assuming that a market-based housing system might facilitate the allocation of housing in a more efficient manner, one might anticipate a gradual adjustment of housing supply and demand in the marketplace over time.

9.3 ASSET MANAGEMENT PRACTICE

The decentralisation of social housing governance is particularly significant for Riga and the large republican cities in Latvia. Riga has close to 30% of the national housing stock, while the other six cities accommodate another 17%. This section provides some insights into asset management practice of Riga municipality—the largest social landlord in the country—with a particular emphasis on rent setting, allocation policies, and sales. The analysis highlights important initiatives that address city-wide priorities in the area of social housing management. Further, it identifies dominant decision-making styles that shape asset management practices today.

Riga municipality is a conceptually appropriate case study for several reasons: i) it is the largest social landlord in the country with the highest share of multi-family housing stock, which creates a particularly challenging context for asset management; ii) the municipal bureaucracy has promoted innovation and has made repeated attempts to reform its institutional systems (rent-setting and allocation) in line with market signals; iii) under the new regime of decentralised governance Riga municipality has the highest fiscal leverage and the ability to carry out its social mandate in a more adequate manner. In sum, the practice of asset management in the city illustrates the most salient features of the transition process in housing as well as responses to major changes in the policy framework.

The analysis draws on qualitative and quantitative methods for data collection. The field work focused on two key elements: i) groups/actors actively involved in housing reforms and the resources they can mobilize to affect change, and ii) perception of stakeholders on results achieved, problems and constraints. Information was collected through key-person interviews, focus group meetings, and primary research during May 2000 and May 2002. Close to 25 stakeholders were interviewed with the objective to elicit responses from all major groups in the public, private and community sector.

9.3.1 Rent Setting

Rents in Riga's social and municipal housing are determined by the Apartment and Services Commission and approved by Riga City Council. Municipal

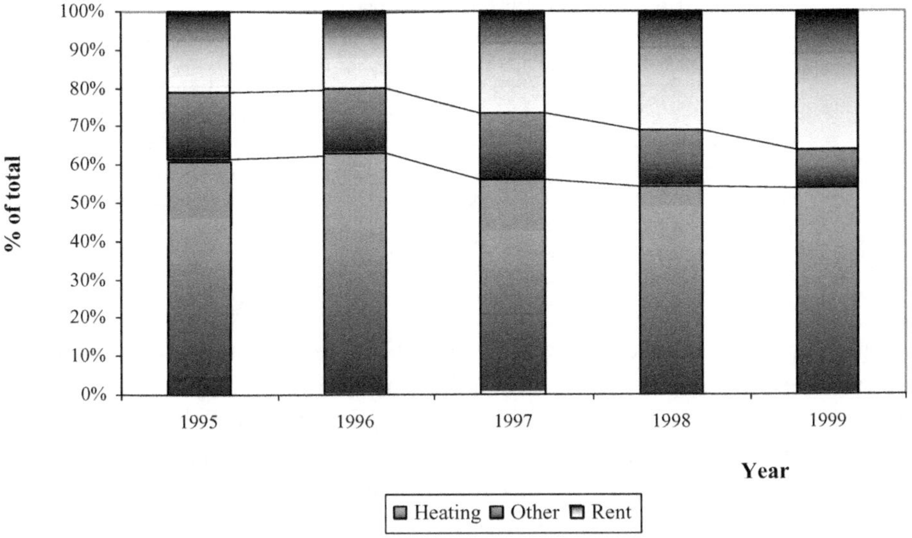

Figure 9.2 Housing costs in municipal housing (1995–1999)
Source: Municipal maintenance company "Mezciems", Riga, May 2000.

maintenance companies charge 0.145 Lats per sq.m. for rent and/or maintenance fees. Although the legislation allows rents to be adjusted in accordance to income, there is no differentiation with respect to housing quality, tenure status or income. Utilities include the cost of heating, hot water and electricity for common areas. Tenants and owners pay an equal amount for the maintenance of the building, which is labelled as 'rent'. Average housing costs (rent together with utilities) constitute 15% of the total household expenditure for the average dual income household.

Figure 9.2 reveals a distorted structure of housing expenditure, with the bulk of the costs consumed by utility payments, rather than fees that sustain the value and the quality of housing assets (Buckley, 2001; Tsenkova, 2000b). Although rents have increased by almost 50% between 1995 and 1999, they still account for less than 35%. Meanwhile utilities, and heating in particular, have reached world prices and have absorbed a disproportionately high share of the household budget. Local experts argue that for some groups housing costs might be as high as 60–80% during the winter months, mostly due to prices of heating and electricity.

Interviews indicate that 8–10% of the households in Riga receive housing allowances with close to 46% of the recipients being single person and

Table 9.5 Housing waiting list, City of Riga

Registration with the municipality	Number of households
Category 1	1,500 households
Category 2	1,000 households
Category 3	4,800 households
Other	1,200 households
TOTAL	8,500 households

Source: Communal Services Department, May 2000.

single parent households (Tsenkova, 2000b).[6] Housing allowances vary between 1–3 Lats per person per month. It is questionable if the support is adequate to cover basic housing needs. In response to affordability problems, Riga City Council has adopted new regulations, which allow tenants in arrears, particularly retired and/or disabled individuals, to receive priority in access to municipal social housing. A new policy allows changes in the status of apartments, provided that households face difficulties in maintaining its rent and utility payment. The 'social status' of the unit will allow rent to be reduced to 0.04 Lats/m^2 and utility payments to be reduced by 75%. These temporary measures eliminate the need to initiate court procedures when debt exceeds 500 Lats. They also allow re-adjustment of the family's financial situation without relocating. Currently 500 apartments have acquired 'social status'; however the need seems to be much greater.

9.3.2 Allocations

Riga has a long history of housing shortages and municipal waiting lists. According to experts from the Housing Policy Division in the Communal Services Department at least 8,500 households are in the newly established 'housing queue', which is approximately 3 per cent of the total population (Table 9.5). The categories and criteria for eligibility are established in the legislation on state and municipal housing assistance. Preference is given to low income households with children, residents with disabilities, and tenants of housing subject to restitution. With non-existent vacancy rates and ongoing privatisation of public housing, Riga municipality is under a lot of pressure to respond to mounting social needs. The new leadership, dominated by social

[6] In Riga direct housing subsidies in the last three years account for 1.5–1.8 per cent of the municipal budget. In addition, the municipality in the last few years has been paying accumulated debt for utilities. Recently housing subsidies have shifted away from general support (available irrespective of income) towards personal, income-tested support. A minimum of 50 Lats per person per month is the established basis for housing allowances.

democrats, has made some commitments to invest in the purchase of existing properties and the expansion of the social housing portfolio. While these efforts might be short lived, and are certainly inadequate given the long waiting list, they demonstrate a growing awareness that investment in social housing is fundamentally important for the welfare of the city and its residents.

Tenant protection in social and municipal housing is still considered to be strong. Despite the changes in the previous Soviet legislation, which provided life-long guarantee of tenant rights and provisions for inheritance of rental housing, tenant eviction for non-payment today is costly for the social landlord, takes at least two years to be enforced, and certainly appears to be politically unpopular. The legislation does not contain provisions for eviction of high-income tenants—the assumption being that these households will take advantage of the 'free right to buy'.

9.3.3 Sales

At the beginning of the housing reforms close to 84% of the housing was publicly owned (Statistic Committee of Republic of Latvia, 1992). Within the general process of ownership transformation, three forms can be discerned: i) restitution/denationalization; ii) privatisation; and iii) accelerated privatisation. The latter implies sale of individual units without the transfer of the building or the land into private ownership. The restitution process had a significant effect on Riga's tenure structure. The process has created a small private rental sector, accounting for approximately 3.75% of the housing stock (Tsenkova, 2000b). While most of the state owned housing has been privatised, interviews with municipal officials indicate that up to 80% of the municipal housing will be sold within the next two years.[7]

Despite the financially attractive conditions, privatisation in Riga has been rather slow, particularly until the end of 1998 (Figure 9.3). Several reasons might explain why this is the case:

- inefficient operation of the municipal authorities and the real property registration system;
- low rents and maintenance costs make the financial situation of a tenant more attractive;

[7] The Law on the Privatization of State and Municipal Apartment Buildings gives the 'right to buy' to existing tenants and members of their family, but also allows privatization by other natural and/or legal persons. The consent of the tenant is required. These provisions, very unique in the context of housing privatization across Central and Eastern Europe, have lead to the development of a market for 'rights to privatize'. In February 2000 close to 25 per cent of the listings of the largest real estate agencies in Riga included these options. Interviews suggest that the asking price for such properties is not really different from the price of outright sales.

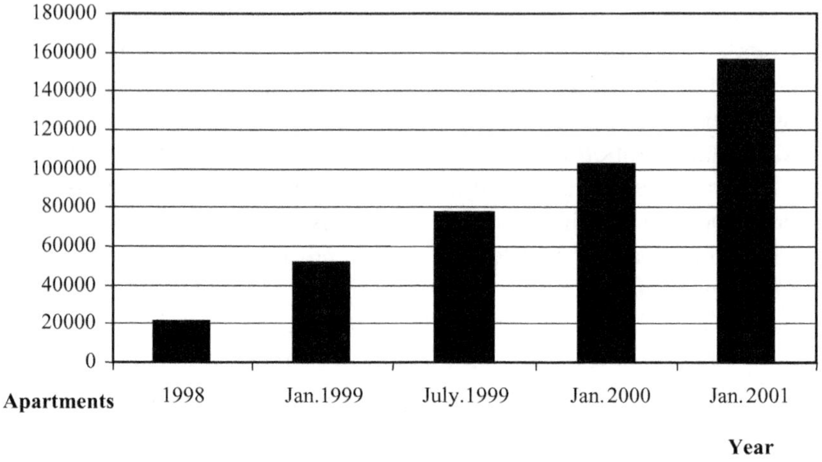

Figure 9.3 Housing privatisation in Riga
Source: Central Statistical Bureau of Latvia, 1999b; Interview Data, May 2001.

– uncertainty about financial and legal implications of home ownership;
– large backlog in maintenance of the public housing combined with generally poor quality of the older housing stock.

Although there does not seem to be any debate about the advantages and disadvantages of housing privatisation, as one the seniour officials pointed out:

> "Privatisation is part of the solutions, but also part of the problem. We learn more about this as we try to implement political decisions. In the housing estates of Riga landlords own the land due to restitution, the municipality owns the buildings and tenants own their flats. Noone really knows how this is going to work."

At the same time, privatisation of state and municipal housing is the largest transfer of housing assets into private hands in Riga's history. A conservative estimate suggests that during 1999 the market value of privatised housing in Riga accounts for 23% of Latvia's GDP (Tsenkova, 2000b). It is in the best interest of all market actors and institutions, as well as the local and the central state, to ensure that the quality and the value of that considerable national wealth are sustained. In addition to its economic dimension, privatisation as a social phenomenon legitimises existing housing inequalities. Some of the residents are entitled to privatise a room in a communal apartment; others have the chance to privatise a good quality, spacious flat. In some cases, new owners acquire a liability, rather than an asset, given the massive need for investment in repair and renovation.

9.3.4 Maintenance and Renewal

Maintenance practices are in the process of fundamental adjustment. The process of change is driven by privatisation, but also by escalating costs for housing services and the lack of systematic approach to the mobilization of funds for routine maintenance as well as capital improvements. Riga's relatively new housing stock consists mostly of apartment housing in multifamily buildings.[8] Close to 45% of the dwellings were built in the last 30 years and as much as 40 per cent of the stock dates back to the period between 1971 and 1989—a result of massive investment in new housing construction during the Soviet era (see Figure 6). According to the availability of amenities, housing quality appears to be high. More than 85% of Riga's housing stock has a central heating system and is connected to the national gas grid. Almost 90% of the housing is equipped with bathroom or shower and more than 99% is connected to piped water. Despite these positive trends, chronic underinvestment in maintenance has left its imprint on the quality of prefabricated housing (Marana, 1998).

During Soviet times, decisions on maintenance and renovation were exclusively driven by technical considerations. Social landlords still employ life-cycle assessment of buildings where different elements need to be replaced in accordance with nationally set standards. While the technical requirements have moved towards harmonization with EU legislation, the major difference is that subsidies are no longer available and the financial difficulties of residents need to be taken into account. To a large extent these new financial and social considerations dominate the decision making process and define what could be accomplished.

There is a growing concern that the City of Riga needs to find alternatives for municipally owned buildings in emergency state of disrepair.[9] The need to deal with repairs and improvement in multifamily housing seems to be another priority. Local programs (e.g. for Mežciems housing district) for replacement of heating installations, insulation and retrofitting are currently underway. In the context of fiscal austerity, however, these measures need to be financed through credits; repayment schemes imply gradual rent increase to account for capital improvements and the cost of servicing the loan. This scenario means policies aiming at rent increase and cost-recovery

[8] Dwellings in Riga are relatively small. The average reidential unit was 48.86 m^2 in 1998, ranging from 32.7 m^2 in dwellings with one room to 91.7 m^2 in dwellings with 4 and more rooms.

[9] Riga appears to have 939 buildings in emergency state of disrepair, which is 11.5 per cent of state and municipally owned housing (Central Statistical Bureau of Latvia, 1999b).

of services, which is politically unpopular. As one of the housing managers stated:

> "Close to 70 percent of the flats on the estate are privatized. People do some improvements but in their own units. No one can agree on anything. We are trying; we get some funding and fix the roof in one building, but then we need more for other renovations. People need to start thinking about ownership and work together to meet these responsibilities . . . "

A particular feature of municipal housing institutions in Riga is the existence of Municipal Housing Maintenance Enterprises (MHMEs). Their main functions are: i) administration of multi-family housing, ii) maintenance and repair of housing stock, and iii) maintenance of public areas and buildings. MHMEs function as municipally owned monopolies within the six districts of Riga.[10] The present combination of privatized housing and Soviet-type monopolistic institutions is considered to be a major problem affecting the establishment of competitive asset management practices.

9.3.5 Decision-Making Process: Considerations, Methods and Strategies

While the central government has rolled back the 'frontier of its intervention', major housing responsibilities were delegated to municipalities without any effort to ensure that they have the means to achieve this end. As a result, the 1990s have seen a range of incremental, and often inconsistent, responses to complex challenges at the municipal level, very much in the spirit of 'muddling through'. As defined by Lindblom (1977) decision-making in terms of 'muddling through' exhibits the following characteristics: i) it proceeds through succession of incremental changes; ii) it involves adjustment and negotiation; iii) it involves trial and error. Policy evolves through complex and reciprocal relations between bureaucrats, politicians, and representatives of interest groups and does not imply fundamentally new approaches (Lindblom and Woodhouse, 1993).

The policy actions undertaken by Riga City Council and individual departments in the last decade have fragmented the policy process, offered concessions to different stakeholders and modified existing policies to the extent possible. The 'muddling through' is reflected in a series of initiatives with major implications for the asset management of social housing. The City of Riga has introduced tenure neutral housing assistance, which is a partial response to affordability problems. Although rents have increased, they do not

[10] In some cities, along with the municipal companies, private entities have gained a large market share—for example in Valmiera (approximately 50%), Ventspils (approximately 40%), Kuldiga (approximately 50%) (COWI, 2002).

ensure cost recovery of housing services and have no relation to quality or location. The result is a simple formula—*low fee no service*. The privatisation of housing has left the largest social landlord in the country with a limited portfolio, often of poor and substandard quality. Further, the 'give away' privatisation policies have stretched the envelope of ownership perhaps too far, creating significant technical, social and financial difficulties for municipal asset management of multifamily housing. With most of the social housing gone, in response to growing numbers on the waiting list and pressures to accommodate socially disadvantaged households, Riga City has invested in the provision of new social housing and the trading of uncompleted residential buildings in exchange for existing dwellings. While these outcomes are the result of central and municipal government policies, they appear to be inconsistent and inherently problematic in a situation of economic recession and high budget deficits.

Within the general policy framework of city-wide control and decision-making two distinct management styles can be discerned in the asset management practice of Riga City: the 'day-to-day asset management' and 'the big bang' solution. The first appears to be the norm in most of the estates and is characterized by a shrinking portfolio, transfer of management to home-owners, and phasing out management responsibilities. The emphasis is on operational management and efforts to balance the budget while avoiding major technical and social problems. Activities are performed very much on an *ad hoc* basis.

The day-to-day asset management on the estate includes two components: i) technical management—monitoring and supervision of local staff and ii) financial management—revenue management, rent and arrears collection (Figure 9.4).

The effort to balance the budget, where local revenue does not exceed local spending, is essential. Municipal maintenance companies no longer receive subsidies from the state or the municipality, which used to be the standard practice under the Soviet regime. While this is a positive step towards more transparent and accountable budgetary processes, chronic underinvestment in maintenance in some of the prefabricated panel buildings has left the social landlords with a lot of unfunded mandates. In addition, managers fulfil other social functions, such as welfare advice and assistance and deal with evictions in the case of rent arrears. These escalating problems, coupled with challenges related to mixed ownership arrangements, require different set of skills and essentially a different breed of housing managers combining technical as well as social skills.

While most estates carry out day-to-day asset management, some have experimented with cyclical maintenance and capital improvements. It is common

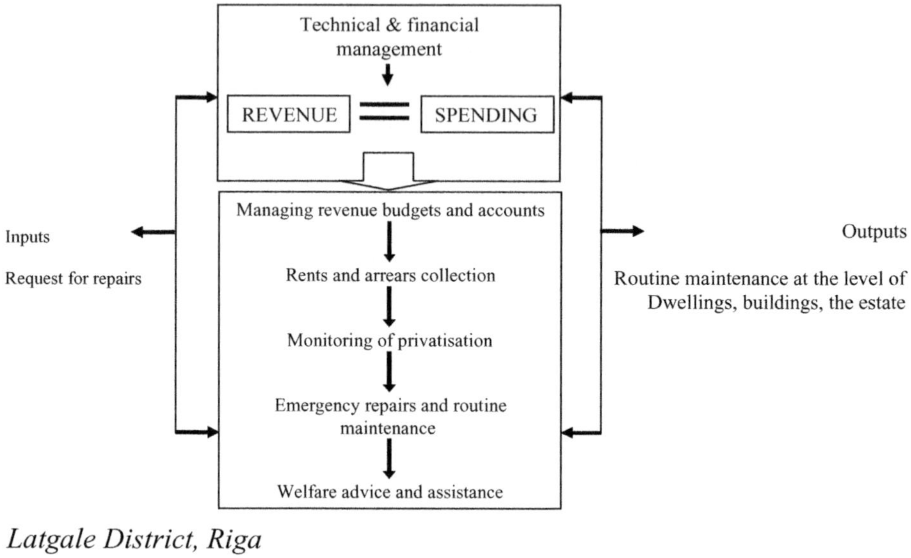

Figure 9.4 Day-to-day asset management

practice to separate the two aspects since the latter requires a more strategic approach to the existing portfolio. The division is also driven by the institutional set up since different municipal departments manage and supervise the process. The main issues are related to: i) what needs to be done; ii) who is responsible and iii) how to finance and carry out the work. This 'big bang' method is considered more proactive, but also expensive. It requires extensive consultation with tenants and owners to mobilise support and funding. Mieziems housing estate uses this approach, which is schematically presented in Figure 9.5.

The practical implementation of this approach so far has delivered limited results. The management has experimented with pilot projects in energy efficient renovation in one or more buildings and/or improvements with smaller credits. The expectations are that the municipality might be able to launch a program that might take these initiatives to a different level. In should be acknowledged that Riga City has made repeated attempts to address its housing problems through a housing reform strategy. Frequent changes in the municipal government and fiscal constraints have made this task particularly difficult. Efforts to improve the efficiency and effectiveness of the municipal bureaucracy and its ability to respond to the housing needs of local residents in a systematic manner essentially address two major priorities:

Figure 9.5 'Big Bang' asset management

- renovation and improvement of housing through energy efficiency measures;
- capacity building programs for homeowners in privatised buildings.

Energy efficiency projects

Interviews with municipal officials and housing managers suggest that the present situation is not conducive to systematic assessment and evaluation of the technical, financial and social factors that will direct investment in renovation and energy efficiency improvements on specific estates. Rather, the approach that appears to be economically feasible involves investment in pilot projects that demonstrate tangible results, which could be promoted to residents and private investors. It is up to the residents and individual estate managers to initiate such projects. While this seems to be crafted in the rhetoric of 'a bottom up approach' to asset management, without systematic support the results so far have been limited.

One successful example is *Riga Energy Initiative*, a demonstration project implemented through German technical assistance. In the first year, the renovated multi-apartment building consumed 51% less energy, which saved 4,094 Lats in heating costs. Other pilot projects have demontrated that reconstruction of a heating substation to allow local control of heating consumption can save up to 20%; installation of meetering devices in individual units—10%, insulation of external doors and windows—20% (COWI, 2002; SCORE Program, 1999).[11] These *ad hoc* activities have gained some popularity in the last two years. The Land and Mortgage Bank is the only institution providing small credits to multifamily buildings for renovation purposes. So far, close to 80% of the loans are dispersed to housing cooperatives and the rest to municipal maintenance companies. Projects are usually quite modest and the average credit is in the range of Lats 10,000.

Capacity building programs

In addition to energy efficiency improvements, Riga City has made a commitment to assist the establishment of home owner's associations in privatised buildings to facilitate the transfer of asset management of common areas. In most of the cases, buildings have mixed ownership, so the association needs to ensure the participation of owners and tenants in the decision-making process. Given the paternalistic culture of asset management in the past, with little to no resident involvement, establishing a new pattern of consultation and consensus building is particularly challenging. "Old habits die very hard. . . . " Despite the privatisation, people's expectations are that the municipality needs to provide services almost free of charge and "take care of things". The municipal maintenance companies are also reluctant to give away their power and share knowledge and competence. The transition in housing governance might take years to become established. The initial efforts have been supported through Canadian technical assistance.

---

[11] Supporting the Co-operative Organisation of Rational Energy Use (SCORE) programme is an initiative of the Dutch government, oriented towards solution of energy saving problems at the user level. Two residential buidngs in Riga were renovated with local partners financiang 2/3 of the total project costs and the remaining 1/3, but not more than Ls 15,000 was a SCORE programme grant.

Latvian-Canadian Housing Cooperation: Institutional Development

The focus of this project is training and capacity building of housing institutions in Riga. A central theme of this policy-oriented professional work is the development of specialized training modules for home owners, the training of trainers and the development of a comprehensive *Manual on Housing Management.* The newly established state and municipal housing agencies have embraced the idea of capacity building for more efficient and effective asset management. *The Manual* has become the best and most comprehensive source of information on legal, financial and technical issues related to housing management, establishment of home owners' associations and a step-by-step guide in energy efficient improvements. So far more than 5,000 copies were disseminated across the country; at least 15 courses were carried out in Riga and other Latvian municipalities delivering training to more than 300 home owners.

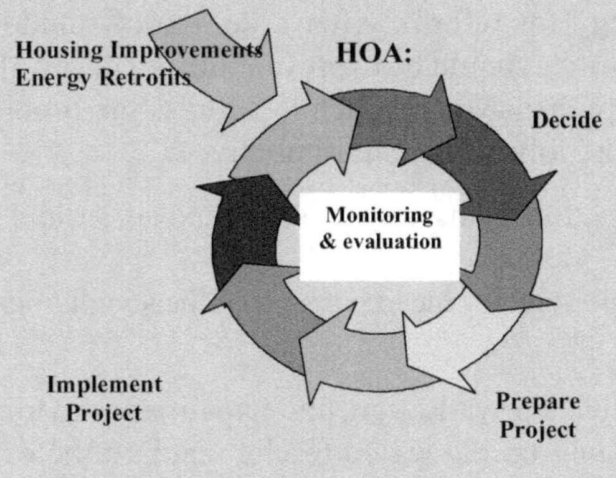

CityInvest (2003)

9.4 CONCLUSION

The 1990s have seen a major change in Latvian housing policies with deregulation of housing markets, privatisation and elimination of explicit state control over the production and allocation of housing. In the context of this wholesale shift away from direct intervention to market-based provision of housing services, this chapter has discussed both the nature of the reforms and their impact on asset management of social housing. Using Riga as a case study, the analysis focused on major housing policies and initiatives and the way these *ad hoc* measures shape housing management strategies.

The policy responses and asset management practices in Riga could be described as 'crisis management', rather than incremental progress towards common goals and objectives. As such, these responses are likely to remain fragmented and incoherent, and not necessarily efficient in economic and social terms. This mode of policy action through trial and error might be a way of institutional learning and a strategy to minimize risk during the transition process. However, in the long run it could simply aggravate the problems in the emerging housing markets and multiply the potential for market failure. While accelerating the restructuring of the social housing sector along market principles is closely linked to the stabilisation of the economy, it also depends on the commitment of municipal and central governments to channel funds into rehabilitation of the existing housing, as well as to undertake further reforms. Possibilities include differentiation of rents, privatisation and competition in the maintenance and management of housing, establishment of financial incentives to ensure cost recovery of housing services and more explicit regulation of owner-occupied multi-family buildings.

At the municipal level, the City of Riga needs to address its housing problems through a coherent strategy for reform. After a decade of 'muddling through' one of the highest priorities should be to prevent further decay of the existing housing stock. A comprehensive approach to address the problems of asset management requires the following major initiatives:

- preparation of municipal action plans for housing renovation and improvement;
- establishment of earmarked funds for the expansion of the social housing portfolio.

Following the completion of voucher privatisation, municipalities need to consolidate ownership in one building envelope and to eliminate, to the extent possible, mixed ownership arrangements. This process of portfolio evaluation has to be supplemented by carefully designed policies dealing with maintenance and management of privatised multifamily buildings. Municipal management obligations should be phased out with the establishment of Home Owners' Associations as a legal entity. Maintenance should become demand-driven, based on owners' obligations for future repair and upkeep of the housing unit and corresponding share of the common space (Tsenkova, 2000b). In the post-privatisation phase municipalities need to focus on more strategic approaches to asset management, as opposed to day-to-day operational activities. Within that context, a systematic overview of rental policy is perhaps overdue. Emphasis needs to be placed on improvement of rent collection procedures, efficient measures to deal with cumulative rent arrears and prudent financial management of limited revenues. Municipalities could introduce rent differentiation to reflect quality and location of the unit and aim at cost recovery of services. This needs to be accompanied by the development of a more efficient and transparent housing subsidy system that targets the needy and abolishes the general subsidies through uniform, low rent policy.

REFERENCES

Buckley, R. (2001) Proposed Housing Project in Latvia. Back-to-office report, Washington D.C., World Bank.

Buckley, R. & Tsenkova, S. (2001) Housing market systems in reforming socialist economies: comparative indicators of performance and policy, European Housing Policy, 1, pp. 257–289.

Central Statistical Bureau of Latvia (1999a) Statistical Yearbook, Riga, Central Statistical Bureau of Latvia.

Central Statistical Bureau of Latvia (1999b) Riga in Figures, Riga, Central Statistical Bureau of Latvia.

Central Statistical Bureau of Latvia (2001a) Privatisation Process in Latvia in 2000, Riga, Central Statistical Bureau of Latvia.

Central Statistical Bureau of Latvia (2001b) Riga in the XX Century, Riga, Central Statistical Bureau of Latvia.

Central Statistical Bureau of Latvia (2002a) Results of the 2000 Population and Housing Census, Riga, Central Statistical Bureau of Latvia.

Central Statistical Bureau of Latvia (2002b) Latvia's Regions 2001, Riga, Central Statistical Bureau of Latvia.

CityInvest (2003) Latvian Canadian housing cooperation: progress report, CIDA, Baltic Initiative Program.

COWI (2002) Energy efficiency and housing studies component Latvia, Riga, Ministry of Finance Technical Unit.

Jaffe A., Turner, B, and Victorin, A. (1995) Property rights and privatisation in the Baltic countries, Copenhagen, Nordic Council of Ministers.

Kemeny, J. (1995) From public housing to the social market: rental policy in comparative perspective, London, Routledge.

Kursis, J. (1999) Housing privatisation in the Baltic states, Lund, Lund University.

Lindblom, C. (1977) Politics and markets, New York, Basic Books.

Lindblom, C. and Woodhouse, E.J. (1993) The policy-making process, third edition, Englewood Cliffs, NJ, Prentice-Hall.

Marana, I. (1998) Housing and emerging problems in Riga, in: Aberg, M. and Peterson, M. (eds.), Baltic Cities, Oslo, Nordic Academic Press.

Markovs V. (1998) Changes in the apartments market of Latvia in 1998, Real Estate in Latvia, 1, pp. 21–25.

Ministry of Economy, Republic of Latvia (1999) Economic development of Latvia report, Riga, Ministry of Economy.

Ministry of Environment Protection and Regional Development (2001) National report on Latvia to the UN conference Istanbul+5, Riga, Ministry of Environment Protection and Regional Development.

Ministry of Finance (2001) Concept for housing lending program in Latvia, Riga, Ministry of Finance.

Social Development Fund (2001) Social assistance in Latvia: statistical analysis, Riga, Ministry of Welfare in Latvia.

Statistical Committee of Republic of Latvia (1992) Riga in Figures 1991, Riga, Statistical Committee of Republic of Latvia.

Tsenkova, S. (2000a) Housing in transition and the transition in housing: the experience of Central and Eastern Europe, Sofia, Kapital Reklama.

Tsenkova, S. (2000b) Riga: housing policy and practice, Riga, Riga City Council.

Tsenkova, S. (2000c) Winds of change: transformation of social housing in CEE, Keynote Paper for the conference Housing in 21st Century—Fragmentation and Reorientation, ENHR, June, 2000, Gävle.

Tsenkova, S. (2002) A concept for a housing monitoring system in Latvia, Riga, CityInvest Inc. & Ministry of Environmental Protection and Regional Development.

10 THE NETHERLANDS

NICO NIEBOER and VINCENT GRUIS

10.1 SOCIAL LANDLORDS IN THE NETHERLANDS

When set within an international context, social rented housing in the Netherlands can be characterised by its relatively large share of the housing stock, the diversity of dwellings within the social rented stock and the variety of tenants, who are not only low-income households (see Table 10.1; van der Flier and Gruis, 2002; van Kempen and Priemus, 2002).

The social rented stock is managed by two kinds of social landlords: housing associations and municipal housing companies. Housing associations are by far the largest providers of social housing. In 2001, there were 620 associations, owning more than 99% of the social housing stock (CFV, 2002). In the last decade, many municipal housing companies were transformed into housing associations by changing their legal status, reducing their number from 213 in 1990 to 23 in 2000.

Because of the small and reducing number of municipal companies, we focus on housing associations. Owning about 35% of the housing stock, they are major players in the Dutch real estate market. They are not-for-profit organisations, which are obliged to operate in the interest of housing, in particular by providing decent, affordable housing to lower-income households. In Table 10.2, some of their characteristics have been summarised.

10.2 NATIONAL CONTEXT FOR ASSET MANAGEMENT

The asset management of social landlords is considerably influenced by the political and economic context in which they operate. After decades of strong central government intervention in the housing market, the Dutch national government policy has embraced the reinforcement of market principles in social housing. As part of this policy, housing associations have gained much more administrative freedom. At the same time, direct financial support for social housing management has been completely withdrawn. This is unique, even within Western Europe (e.g. Boelhouwer, 1997, 1999). The new policy context has set considerable challenges for the asset management of Dutch social landlords. Being transformed from operational, task-oriented

Vincent Gruis and Nico Nieboer (eds.),
Asset Management in the Social Rented Sector, 185–208.

Table 10.1 Division of the housing stock by tenure in the Netherlands (1 January 2001)

	Social rented	Private rented	Owner-occupied
Percentage of total housing stock	35%	11%	53%

Source: MVROM (2002)

organisations towards 'social entrepreneurs', they have to operate in a more strategic, market-oriented way. As a result, there has been wide interest among associations in methods and instruments to support asset management in a more systematic and business-like manner.

10.2.1 Regulations

General administrative regulations

The legal rights and obligations of Dutch housing associations are formulated in the Social Housing Management Decree ("Besluit Beheer Sociale Huursector"—BBSH). This decree stipulates that all the activities of housing associations have to be in the interest of housing. Housing associations must give priority to accommodating households with a weak position on the housing market (mainly lower-income households). However, they are allowed to provide dwellings for other target groups. They are also allowed to deliver high-rent or owner-occupied housing. As a consequence, Dutch housing associations are often typified as 'hybrid' organisations, which carry out public tasks, but are independent, private organisations, having market-driven objectives as well (Priemus, 2001, pp. 247–249). The BBSH formulates the obligations of housing associations in the form of general 'fields of performance': accommodation of target groups, preservation of the quality of dwellings and their environment, consultation of tenants, securing the financial continuity and providing housing and care arrangements.

Table 10.2 Some characteristics of housing associations in the Netherlands (2001)

Characteristic	Number/Percentage
Number of dwellings	2.3 million
Number of landlords	620
Dwellings per landlord (average)	3,800
% of total housing stock	35%
% of social rented stock	99%

Source: CFV (2002), MVROM (2002).

Table 10.3 Rent policy for the period 2002–2005

Rent as percentage of maximum eligible rent	Maximum annual rent increase
70–100%	inflation
60–70%	inflation + 1%
50–60%	inflation + 1,5%
<50%	inflation + 2%

Source: Remkes (2002)

Rent setting

For the major part of the dwellings in the Netherlands (expensive dwellings are excepted) the so-called Housing Valuation System ("Woningwaarderings-systeem"—WWS) is in force. On the basis of this valuation, a maximum eligible rent can be calculated. In principle, rents can only be changed on July 1st of each year. The government decides each year the maximum allowed rent increase. For housing associations, the government also determines each year a maximum rent increase on corporate level. The latter regulation is in force since 1993. Before, the government determined the annual rent increase on individual level and housing associations were not allowed to vary the annual rent increase per dwelling themselves.

Since July 2002, new national rent regulations are in force, which substantially confine the policy freedom of housing associations that they got in 1993. The former State Secretary of Housing implemented a mid-term rent policy for the period until 2005, which links the maximum rent increases to the inflation rate and the maximum eligible rent according to the Housing Valuation System (see Table 10.3). Furthermore, the government sets a maximum to the rent increase of the total stock of a housing association at inflation plus 0.4%. Based on the difference between the current and the maximum rent, the maximum rent increase for individual dwellings varies between the inflation rate and the same rate plus 2%. In 2005 the introduction of a modified Housing Valuation System is planned. The slow progress of the negotiations between the landlords' and the tenants' national representatives, however, has made the introduction of this system highly uncertain.

Allocations

Housing associations are rather free in the way of allocation, but within certain government restrictions. A national restriction is that relatively cheap homes must be allocated to low-income households. Sitting tenants, however, cannot be evicted because of (increased) income.

Many housing associations work together with the local government to create and manage distribution systems. Common decisions are made on allocation criteria, for example, relating to household income, size, or age (Kullberg, 2002).

Sales

Housing associations are allowed to sell dwellings. In fact, they are encouraged to do so, since the promotion of home-ownership is a prominent issue of Dutch housing policy. Since March 2002, only plans to sell dwellings to others than individual households (e.g. to private landlords) and plans for sale which do not fit into performance agreements with the local authorities have to be reported. In principle, housing associations must sell their dwellings against a minimum of 90% of the market value. However, if the buying household makes use of the Purchasing Grant—established as part of the Act on the Promotion of Home-Ownership ("Wet Bevordering Eigen-Woningbezit"—BEW)—associations may give reductions to 20% for new households and 30% for actual tenants. Additional reductions are possible if the development of value of the home after the sale is shared between the buyer and the housing association. The BEW, however, is hardly used in practice, mainly because it can only be applied in a small market of relatively cheap dwellings.

Maintenance and renewal

The regulations for the technical quality of the housing stock (including social rented dwellings) have been laid down in the Building Decree ("Bouwbesluit"). This decree states minimum (technical) requirements on, among others, construction, safety, energy consumption and health. The Rent Act for Housing ("Huurprijzenwet Woonruimte"—HPW) mentions several additional technical requirements that, if not met, give tenants the right to pay a substantially lower rent. The BBSH states no specific, additional regulations for the quality of social rented dwellings.

10.2.2 Supervision

Housing associations are supervised on the basis of retrospective accountability. Their efforts in the interest of housing have to be reported to the Minister of Housing. Their financial position is monitored by the Central Housing Fund ("Centraal Fonds voor de Volkshuisvesting"—CFV). The BBSH contains prescriptions for the annual reports. The government may impose sanctions if an association performs poorly or in conflict with regulations, such as a directive (to undo or to perform a certain activity) or the appointment of a temporary supervisor. Furthermore, housing associations are encouraged to draw up 'performance agreements' with the local authorities. In these agreements, they specify their (public) tasks.

In addition to the external supervision by the government, associations have developed an internal supervision structure as an additional means to ensure that they will operate within the interest of social housing in a reliable way. Within the social rented sector an 'audit-system' has been set up to review the performances of housing associations as well as a Business Code with 'rules of conduct' for the members of the umbrella organisation Aedes. Within individual associations, management is supervised by a board of 'supervisors' or 'commissioners'.

10.2.3 Financial Support

As stated in the introduction, direct financial support for housing associations has been diminished. During the 1990s, brick-and-mortar subsidies, exploitation grants and central government loans have been abolished. Indirect financial support still exists through:

- some tax benefits, although these have been announced to be abolished;
- housing grants to individual households (IHS);
- the backing up of a guarantee structure for the social rented sector (see below).

In principle, the Dutch social rented sector is expected to act as a 'revolving fund' in which financial returns are reinvested in the interest of housing. Furthermore, there has been established a financial guarantee structure which provides associations with good access to the capital market. Associations' loans can be guaranteed by the national Social Housing Guarantee Fund ("Waarborgfonds Sociale Woningbouw"—WSW), which is filled by fees from the associations and backed up by the government. Associations that are no longer able to secure their financial viability can apply for financial support from the Central Housing Fund (see e.g. Boelhouwer, 1997). As part of the revolving fund, 'rich' associations with few tasks at hand are expected to help 'poor' associations with many tasks in their locality, for example by providing loans against low interest rates. However, this 'matching' of means and tasks has been a marginal phenomenon until now. This has led to an ongoing debate whether obligatory measures are necessary and desirable. So far, these discussions have not resulted in such measures.

On average, Dutch housing associations are in a relatively good financial position (see e.g. Priemus, 2001), although there are considerable differences between individual landlords. For the next ten years, a further improvement of the financial situation is predicted (CFV, 2002).

Table 10.4 Housing shortage in the Netherlands

Year	Number of dwellings	Percentage of total stock
1994	134,000	2.1%
1998	90,000	1.4%
2000	93,000	1.4%
2005 (expected)	186,000	2.6%
2010 (expected)	181,000	2.5%

Source: den Otter and Heida (2002).
Note: the expansion of the housing stock in this forecast is based on MVROM (2001).

10.2.4 Housing Market

In general, the Dutch housing market is acknowledged to having neared a quantitative equilibrium between supply and demand. However, according to the latest prognoses, the housing shortage may be expected to rise again in the next decade, mainly due to a stagnation in the developments of new dwellings (see Table 10.4). Furthermore, there are considerable regional differences. In the 1990s, in some regions there have been surpluses resulting in a significant number of vacancies, whereas in other regions the pressure on the housing market has remained high.

There are substantial differences between the actual composition of the housing stock and the demanded quality, price and tenure. As we can see in Figure 1, the largest surpluses concern the cheaper dwellings, and in particular the cheaper rented, multi-family dwellings—of which most are in the possession of housing associations.

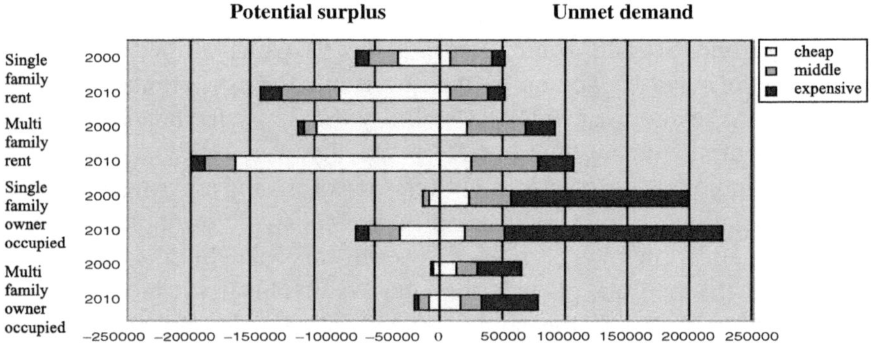

Figure 10.1 Forecast of discrepancies between supply and demand
Source: MVROM (2000a).
Note: ‘potential surplus’ does not equal vacancies, nor does ‘unmet demand’ equal a quantitative housing shortage. They reflect mismatches between the actual dwelling and the desired dwelling (MVROM, 2000a).

On the macro level, the surplus of cheaper rented apartments implies that associations are facing a reducing popularity of a significant part of their housing stock. In general, however, this is not reflected in lettability problems because of the general housing shortage and the sharp increase of prices in the owner-occupied sector in recent years.

10.3 ASSET MANAGEMENT PRACTICE

In this section, we give an overview of the asset management practice in the Netherlands. As in the previous section, we do this by following the activities that are part of asset management. Furthermore, we will reflect on the decision making process and the instruments used to support asset management decisions. Our description is of a general nature, based on a literature review, interviews with eleven housing associations and examples from supposed 'front-runners' among Dutch associations.

10.3.1 Rent Setting

As we have stated in the previous section, housing associations gained considerable freedom in their rent policy in 1993, which enabled them to adopt a more market-oriented approach. Since then, many associations have differentiated the yearly rent increase for (groups of) dwellings. In doing so, they generally work with 'target rents', which are based on the quality and market position of the dwellings. Most associations fix the target rent at a percentage of the maximum eligible rent according to the Housing Valuation System. Many associations correct these target rents (or percentages) on the basis of (indicators for) the popularity of the dwellings (to reflect the market position) (e.g. Thomsen et al., 2001).

The market orientation of the housing associations' rent policy as is reflected in the results of surveys conducted by the research institutes OTB in 1996 (Kersloot, 1999) and RIGO in 1999 (Marsman and Smit, 1999). In the OTB survey, the most frequent reason that housing associations mention for their rent increases concerns the price-quality rate of the dwellings; the second most frequent reason concerns the affordability and/or price-quality rate of other dwellings. In the RIGO survey, the most frequent reasons concern the local housing market and the financial position of the associations.

Associations in the Netherlands are experimenting with alternative rent contracts, to bind their tenants, to give them more possibilities of choice, and to give them more security about their rent payments. Examples are:

- 'fixed rent' contracts in which the rent is initially set higher but will not increase in the next five or ten years;
- 'fixed rent increase' contracts in which, for example, the initial rent is set lower, but the rent increase is set higher in the next five or ten years.

Rent Policy Delftwonen

Delftwonen determines the target rent in relation to the quality and market-position of the dwellings. The target rent is based on a percentage of the maximum eligible rent according to the WWS. This percentage is based on an additional valuation of locations and dwelling types, according to the following table:

Market position of location[1]	Single family dwelling	Ground-floor dwelling (in multi family block)	Multi family block with doorway and elevator	Gallery apartment with elevator	Multi family block without elevator	Single person dwelling (in multi family block)	Dwelling for elderly
A	90%	90%	90%	90%	90%	90%	85%
B	80%	80%	80%	80%	80%	80%	85%
C	80%	80%	80%	75%	75%	75%	85%
D	80%	80%	80%	75%	75%	70%	85%

The yearly rent increase depends on the difference between the current rent and the target rent, taking into account the government regulations shown in Table 3. At turnover the rents are increased to the target rent at once. Before implementing this rent policy, Delftwonen has checked if enough affordable dwellings remain to comply with the need according to the local market prognoses and Delftwonen's targets for its contribution in meeting this need. For the whole stock, Delftwonen strives towards an average rent increase equal to inflation.

10.3.2 Allocations

In the Netherlands, two basic systems are used to allocate social rented dwellings. Up till the 1990s, dwellings were allocated according to the 'distribution model ' or 'waiting list model'. People subscribed to municipal waiting lists and were ranked according to waiting time and specific needs. After a successful experiment in the city of Delft in 1990, choice based lettings became common practice. In this so-called 'Delft model', eligible home hunters can directly respond to individual vacancies that are publicly advertised, for instance in a local newspaper or magazine. Social priorities are shaped through an open queuing system with clear ranking criteria. In order to distribute the stock according to people's needs and means, conditional criteria are applied. The (conditional and ranking) criteria are usually based on (combinations of) household size, income level and age. Also, a distinction is often made between 'transfers' and 'starters', who are ranked on the basis of respectively the number of years in their current residence and the number of years on a waiting list or their age (Kullberg, 2002).

Allocation Dwellings Delftwonen

Delftwonen participates in the 'Delft model' for the region Haaglanden (the region around the city The Hague), in which almost all the associations' dwellings in this region are offered. In this model, conditional criteria are attached to:

- household size, in relation to dwelling size;
- age, for dwellings specifically meant for young people or elderly;
- household income;
- household 'market career': starting or transferring household.

Following an experiment in the period 2000–2001, dwellings are also labelled on the basis of the desired 'chance of success' that groups of households have on acquiring a social rented dwelling. For the whole region it has been agreed that the chance of success for households with minimum wages has to be higher than other target-group households. The latter must in turn have a higher chance of success than non-target-group households. Within Delft is has also been agreed that the chance of success for starters and transferring households have to be equal. Within the group of starters the chance of success has to be approximately the same for each age. In practice, this would mean that more (and also bigger) dwellings have to be labelled for starters than in the past.

Starters are ranked on the basis of their searching period. Transferring households are ranked on the basis of the period they live in their current dwelling. Furthermore, priority is given to:

- households that move from a cheap dwelling within the region to an expensive dwelling;
- people older than 65 years who want to move within their neighbourhood;
- people older than 55 years who want to move within their neighbourhood and leave a cheap dwelling.

Sources: van der Zon (2002) and www.woonnet-haaglanden.nl.

Table 10.5 Reasons for sale given by housing associations for dwellings sold in 1995

Reason for sale	Of importance (%)	Of no importance (%)
Expected vacancies	73	27
Dwellings became too expensive for target group	72	28
To improve financial position	66	34
To use proceedings for acquiring new dwellings for target group	64	36
To prevent high turn-over rate	57	43
Market value exceeds value at (social) rent	57	43
To increase differentiation of (neighbourhood) population	55	45
At tenants' request	47	53
To use proceedings for acquiring new popular dwellings	46	54
To use proceedings for improvements	40	60
To use proceedings for maintenance	32	68
On the basis of agreements with local authority	20	80

Source: survey by OTB Research Institute in 1996, reported by Kersloot (1999).
Note: the percentages are weighed on the basis of the size of the associations' stock.

10.3.3 Sales

Because of the increased administrative and financial independence of the social sector and because of market developments, sale of dwellings has become a prominent activity of many housing associations. The amount of sold dwellings grew from 2,000 in 1990 to over 20,000 a year in the late 1990s. According to the Ministry of Housing 'mainly the more expensive dwellings are being sold and the cheaper stock seems to be maintained as much as possible for housing the target group' (MVROM, 2000b, p. 11). The reasons that associations have stated for selling their dwellings (see Table 10.5) shows that their decisions are based on a mixture of market, financial and social considerations. High on the list are 'the use of proceedings from sale to acquire new dwellings for target group' and 'dwellings became too expensive for target group'.

Sale Policy Delftwonen

In determining which dwellings can be sold, Delftwonen takes the following objectives into account:

- the sale fits into the renewal strategy for the neighbourhood and leads to a more differentiated neighbourhood with a higher quality;

- the sale facilitates the housing career of the current households in a neighbourhood or keeps them within the neighbourhood;
- the sale facilitates the financing of other (unprofitable) investments in a neighbourhood;
- the sale fulfils a significant role in meeting demand and reduces problems of lettability.

'Hard' criteria for sale are:

- dwellings are only sold in a good technical condition;
- there has to remain a sufficient amount of social rented stock to supply the target group with a diversity of dwellings in diverse neighbourhoods;
- dwellings will not be sold if they are located in strategic areas, where redevelopment is expected to take place in the (near) future.

Dwelling which have been allocated for sale as part of Delftwonen's strategic asset management plan are first tested against these criteria, before they are actually offered for sale.

In addition to 'regular' sales, associations have developed alternative forms of sale, which are meant to make purchase affordable for a larger part of their target group:

- sale of the interior, while the exterior of the dwelling remains to be rented from the association;
- sale under conditions for resale (when the initial buyer wants to sell, he first has to offer the dwelling to the association) and sharing the value development of the property at the moment of resale (e.g. on fifty-fifty basis).

A new phenomenon is the 'tenure-neutral' dwelling. 'Tenure-neutral' in this sense means that the choice between rent, 'traditional' sale or alternative forms of sale is not made by the landlord, but is left to the present or new inhabitant. In most cases, it concerns sale under conditions of resale by the housing association. Nowadays, some housing associations have serious plans to offer a part of their stock as 'tenure-neutral' dwellings.

10.3.4 Maintenance and Renewal

Until recently, the management of the housing stock was mostly a routine activity, carried out at an operational level in the organisation. Nowadays, housing associations realise that maintenance and investments ought to be based not just on technical, but also on market considerations. In this context, several client-oriented maintenance policy options are considered, for example:

- linking the level of equipment to the target group (e.g. dwellings that are targeted for higher incomes are equipped in a more luxurious manner);
- allowing tenants to make (minor) adjustments in their dwellings;
- involving tenants in maintenance decisions, for example by allowing them to choose between a series of kitchens or bathrooms.

Straub (1997) has found that almost all housing associations carry out planned maintenance on the basis of a long-term, computerised maintenance planning. Inspectors register defects and the extent of these defects during inspection. Most housing associations also indicate the technical urgency of planned maintenance activities. A growing number of housing associations register the maintenance condition of building components in quantitative terms (e.g. using values on a six-point scale). By adopting a condition-dependent approach to maintenance, housing associations can exercise control over the desired maintenance performance levels and costs. In general, housing associations discern a standard, 'normal' level of maintenance, a 'substandard' level (for instance if demolition or refurbishment is foreseen in the near future), and a higher level. The latter option can be useful to improve the market position of the property (Straub, 2002).

Another, related issue in the social housing sector is the way to develop a more client-oriented maintenance policy, for example by:

- linking the level of equipment to the target group (e.g. dwellings that are targeted for higher incomes are equipped in a more luxurious manner);
- allowing tenants to make (minor) adjustments in their dwellings;
- involving tenants in maintenance decisions, for example by having them choose a kitchen or bathroom within certain constraints.

Renewals are induced by negative developments in the market position and the (technical) quality of the dwellings and the occurrence of social problems in neighbourhoods. Unlike maintenance, renewals are usually treated as individual 'projects' and therefore not part of the going-concern operations. Although associations are expected to invest substantially in the improvement of their post-war housing stock, the statistics show that this has not yet been taken place at a large scale (van der Flier and Thomsen, 1998).

Maintenance Policy Delftwonen

Delftwonen's maintenance policy is based on the technical need on the one hand, and on the 'label' for the quality level which has been attached to an estate on the other. The quality label is determined as part of the strategic asset management plan. Four labels are used: 'preserve', 'sober', 'standard' and 'luxurious'. The labels are based on:

- the rent level and the 'lifestyle' of the target group for the dwelling;
- the current quality;
- specific strategic plans for an estate (for example renewal, sale or demolishment).

The level of maintenance and equipment of most dwellings of Delftwonen are classified as 'standard'. The label 'sober' is appointed to dwellings that cannot reach the 'standard' level, unless refurbishment takes place. This level is also applied to a number of dwellings, of which the tenants prefer a lower quality and a lower rent in exchange. The label 'preserve' is only applied to dwellings that will be renewed or demolished within the next five years. 'Luxurious' is meant for dwellings that, due to their rent level, are only affordable for the middle and higher income groups. With this category, Delftwonen aims to diversify its stock and to generate financial means).

10.3.5 Decision-Making: Considerations, Methods and Strategies

Considerations

Apart from developments within the maintenance policy of housing associations, there is a shift in the relative importance of maintenance and other technical issues. Until the 1980s, technical considerations were dominant in investment decisions in the existing stock. Generally speaking, the technical state of a dwelling or a building determined if improvements were made. In a survey held in 1997, housing associations were asked to indicate the importance of several factors in their stock management, such as technical quality, equipment (e.g. central heating, insulation), availability of subsidies, tenant preferences, lettability, exploitation benefits and property value. Among these factors, lettability was mentioned to be the most important one, followed by equipment (Straub, 1997, p. 36, see Table 10.6).

Nowadays, periodical reporting on present lettability and vacancies is common use (see e.g. Nieboer and Gruis, 2002). Especially landlords in areas of

Table 10.6 Factors influencing the asset management according to Dutch associations

Considerations in housing stock management	Determinative or largely determinative (%)	Not or little determinative (%)
Market perspective (lettability)	95	5
Equipment	86	14
Social problems	74	26
Technical quality	70	30
Financial return	69	31
Type of dwelling	68	32
Income stream value	65	35
Net worth (financial strength association)	62	38
Sitting tenants preferences	54	46
Quality urban structure neighbourhood	45	55
Administrative value	33	67
Building year	24	76
Building regulations	18	82
Availability of grants	17	83
Quality of the environment	17	83
Prestige (publicity)	13	87

Source: Straub (1997).
Note: the emphasis in Straub's survey results on factors relating to quality (in contrast to e.g. affordability and availability) may be partly explained by the focus on technical housing management and investments in the rest of his survey.

low demand still seem to be very keen on lettability developments (frequent reports on lettability etc., direct action in terms of allocation, rent or even sale), despite little lettability problems.

Mainly due to the increased importance of market factors in social housing, several marketing concepts have been developed, in which present and potential tenants are categorised according to their 'life-style'. These concepts can be quite different from traditional categorisations of households that are based on age, income and household size.

Use of Life-Style in Asset Management

Quite remarkable is a division, developed by a private company (named SmartAgent Company), based on individualism and compliance with traditional norms and values in society. On the basis of this classification, the following categories are formed (cf. Hagen, 2001, pp. 9–10):

– the 'independent' (not a newspaper, but referring to individualistic and non-compliant people);
– the 'dynamic individualistic' (compliant and, of course, individualistic);
– the 'community persons' (non-individualistic, not specifically compliant or non-compliant);
– 'silent luxury' (non-individualistic and compliant);
– 'withdrawers' (non-individualistic and non-compliant);
– the 'embedded' (very group oriented and non-individualistic, not specifically compliant or non-compliant).

These life-styles have also been used by Delftwonen to determine for which people an estate is (potentially) attractive

For Dutch housing associations, financial considerations may be expected to have gained importance during the last decade, since government support has diminished. However, results from our research (Nieboer and Gruis, 2002) indicate that it is very uncommon for housing associations to use goals for return, risk and value development. Although the research shows some first signs of thinking in these terms, finance plays only a minor role in the first stages of decision making about stock development. As for the interviewed housing associations, the role of finance has been found in feasibility studies *after* the formulation of investment proposals rather than *during* the formulation of it.

From our research, we cannot say if this applies to all social landlords in the Netherlands. On the one hand, the modest role of financial considerations is supported by the fact that in 2000, 7% of the investment costs by Dutch housing associations was not covered by the net present value of additional proceedings generated by these investments (Aedes, 2001). Furthermore, the net present value for newly built dwellings shows an average negative value of about € 15.000 per dwelling (CFV, 2001). Landlords settle these financial shortages through their own financial reserves, the sale of (extra) dwellings and the development of profitable (owner-occupied) dwellings (de Kam, 2001). On the other hand, some surveys show that housing associations attach a lot of importance to financial considerations (e.g. Marsman and Smit, 1999; Straub, 1997; Kersloot, 1999). From some landlords outside this research we know that a desired solvency level can considerably affect their asset management, as in most of these cases sale of dwellings is necessary.

A remarkable phenomenon is the development of a financial real estate index for Dutch housing associations, named the Aedex/IPD Social Housing

Index, which is similar to the IPD real estate index that is already used in the commercial real estate sector. This development reflects the growing importance of financial considerations in housing associations' policies.

Next to transparent measurements of financial return, there is an increasing need of measurements of 'social return'. This is more or less a consequence of political debates about the efficiency and effectivity of social landlords. Because social return is hard to define, there is, until now, little progress in making social return measurable. We have found no other output objectives than keeping a minimum number of properties for the lower income households. Social considerations, nevertheless, form an essential factor in decisions about stock development. The considerations are, however, in the minds of decision-makers and are not made explicit in output terms.

Sustainability is another factor of growing importance, especially since the early 1990s. In 1998, the umbrella organisation for Dutch housing associations drew up a voluntary agreement with the Ministry of Housing to strive towards sustainable building and greater energy efficiency in new building and housing management. However, despite various experiments, sustainability is not an important factor in asset management of Dutch housing associations (Sunikka and Boon, 2002).

Top-down strategies

The reduction of government support in social housing has increased the need among housing associations of formulating objectives for the development of the portfolio. In specialist literature there is a vast amount of articles on methods and techniques to support policy development and decision making.

Nevertheless, little can be found about portfolio strategies and, equally important, the translation of these strategies to exploitation measures on the estate level. Recent research (Nieboer and Gruis, 2002) indicates that this is not only the situation in literature, but also in reality. Most goals and objectives are rather general (for instance: 'quality standards must be raised, 'to meet the growing demand of the elderly' etc.) and not quantified. Despite various initiatives to define 'top-down' guidelines for the development of the housing stock, a 'translation' of these guidelines into exploitation measures is missing. As a consequence, investment policies are mainly determined on the basis of information on estate level; goals and guidelines on portfolio level play a subordinate role in asset management.

Structuring the decision making process

As a consequence of the tendency towards a more business-like management, several process models have been set up to develop this kind of policy in a systematic way (Nieboer and Gruis, 2001). Basically, these models consist

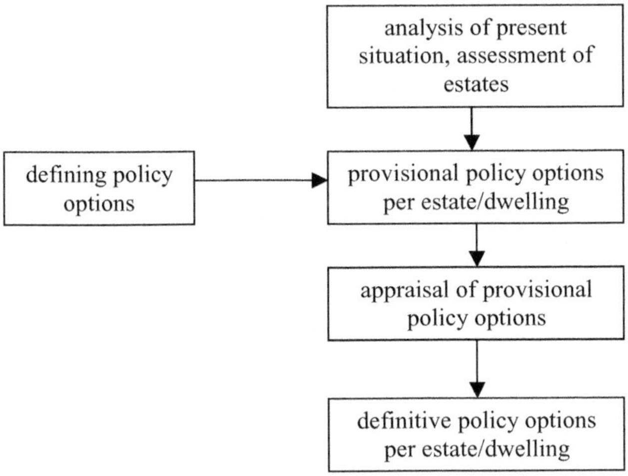

Figure 10.2 General outline of existing models for strategic asset management.

of a stepwise description of the actions to be taken to formulate stock policy and provide individual landlords with a framework for policy making. The models are specifically designed for the social sector because, unlike most private landlords, Dutch social landlords have a geographically concentrated stock and must meet a number of legally imposed public tasks. This implies that, compared to commercial landlords, social landlords must pay more attention to social and political considerations, as well as factors concerning the neighbourhoods in which the housing stock is concentrated.

Although a number of models have been made to develop strategic stock management in a systematic way, only a few models have been published. There are two probable reasons for this. First, many models are constantly changing and are adapted to every new situation, according to specific (local) circumstances. Second, several models have been developed by consultants, who are not always willing to publish their models because of commercial reasons.

The models that have been published have a common-sense sequence of analysis, formulating provisional policy options, testing (ex ante evaluating) these options (option appraisal), and formulating definitive options (see Figure 10.2). These strategies consist of a series of planned measures towards the housing stock. Some models also pay explicit attention to the policy principles of the landlord or to the implementation of the planned measures (see Van Leent and Van Vliet, 1992; Van Ginneken and Van der Laag, 1992; Van Vliet, 1993; Van den Broeke, 1995, 1998; Nieboer et al., 2001).

Decision making process De Combinate

The process followed at housing association De Combinate (11,000 homes, all in Rotterdam) can be divided into seven steps:

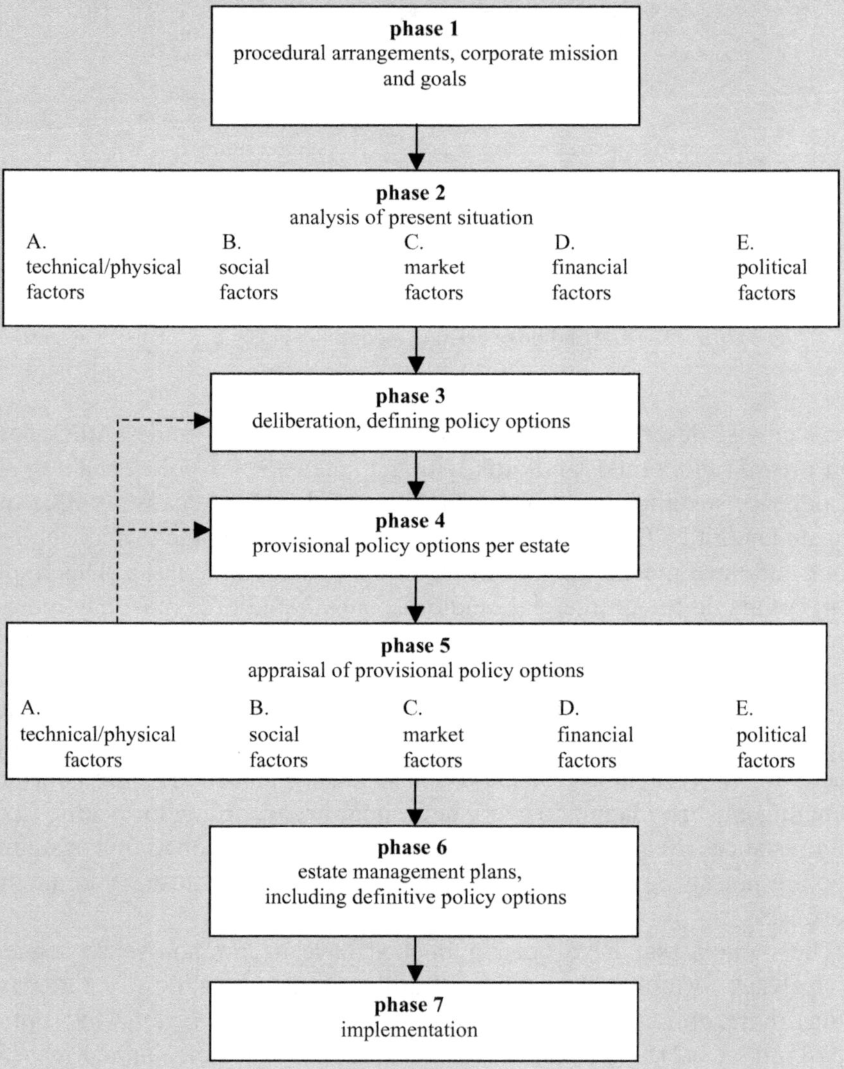

It is unknown how many housing associations follow the structure of the process models mentioned above. From our experience it seems plausible that this way of developing asset management is highly brought upon by external advisors and is (still) not a regular, firmly embedded procedure at the housing associations themselves. Besides, several housing associations have started separate projects for neighbourhoods in which regeneration of the housing stock will take place. The structure of decision making in these projects is rather a product of relatively unstructured negotiations with external parties (local government, other landlords etc.) than an internally structured approach.

Methods for option assessment

Several methods have been developed to connect data on individual estates or properties to policy options. In a simplified form, all these methods can generally be described as a set of 'if... then' statements which take form in portfolio analysis and decision trees.

Based on theory on strategic business planning (e.g. Ansoff, 1984), it has been suggested that housing associations can use a marketing portfolio analysis of the market position in support of management decisions (e.g. van Vliet, 1993; van den Broeke, 1998). In such an analysis, estates are assessed on the basis of their current market position (e.g. on the basis of vacancy and turnover rates) and their future market prospects (e.g. on the basis of housing market research). The outcome of this analysis may be translated directly into suggestions for basic strategies to follow (e.g. 'grow' or 'cherish' when market position and prospects are good and 'divest' when market position and prospects are bad). In this way, portfolio analysis may contribute to a systematic approach to formulate strategies for the housing stock (cf. van der Flier and Gruis, 2002).

In recent research among eleven housing associations (Nieboer and Gruis, 2002) only one housing association has been found that uses a kind of marketing portfolio analysis as part of its stock management. Three other housing associations use a modified version of portfolio analysis, for instance with lettability and price-quality ratio as the principal elements. The role of the portfolio analysis in asset management decisions is, however, modest. This gives the impression that this technique is not widely used in practice and, if used, does not play an important role. This may be partly explained by the fact that portfolio analyses have been designed initially for commercial purposes and, for this reason, cannot be applied directly to social housing management.

In addition, insight in *future* market situations is less developed. Forecasts of market situations, if there are any, do mostly not contain information on the appropriate geographical scale, thus leaving a lot of uncertainty about the future market situation of a landlord's portfolio. Moreover, several landlords have indicated that they regard the market too dynamic to rely on these forecasts, even if these would specifically be made for their property. For this reason, future market prospects are more based on past experiences and subjective knowledge than on systematic calculations. This, in turn, may partly explain the rare application of marketing portfolio analyses.

Portfolio Analysis De Combinatie

In the development of strategic asset management at housing association De Combinatie, a marketing portfolio analysis is used to select potential policy options per estate. These policy options are structured in 'labels' that indicate tenure (continuing exploitation, sale or demolition), eventual change of target group (reallocation or not) and technical quality improvement (small improvement, upgrading, refurbishment). The portfolio analysis is illustrated in the following table. The selected policy options are confronted with estate-specific bottom up information about each estate. In this way, provisional policy options are chosen, which will be tested against a range of criteria (costs, capacity etc.) in the next phase of policy making.

Market perspective lettability	Bad (shrinking market)	Moderate (stable market)	Good (growing market)
good	consolidation small improvement upgrading	consolidation upgrading	consolidation upgrading
moderate	reallocation upgrading upgrading/ reallocation sale demolition	consolidation small improvement upgrading	consolidation upgrading
bad	sale demolition	reallocation upgrading upgrading/ reallocation sale demolition	consolidation small improvement upgrading

Decision Support Model Delftwonen

Delftwonen uses a decision tree, based on a 'classic' portfolio analysis, in which strategies are based on assessments of future market perspective, current lettability and economic opportunity costs (measured as the ratio between the Net Present Value under continued social rent and the market value):

Market perspective	Lettability	Economic opportunity costs	General strategy
No risk	Good	Low	Grow
		High	Milk
	Bad	Low	Reinforce
		High	Improve marketing, retreat
Risk	Good	Low	Maintain
		High	Maintain (for now)
	Bad	Low	Reinforce, retreat
		High	Improve marketing, retreat

The general strategies in this decision tree are, shortly described, as follows:

- Improve marketing: improve lettability, but without increasing economic loss, for example by changing allocation criteria and being more active in marketing. If investments in quality are deemed necessary, they should be financed by the sale of dwellings in the same product-group;
- Grow: develop more of these products and cherish the ones in the current stock;
- Maintain: continue current policy for these products;
- Milk: improve financial return by increasing rents;
- Retreat: sell or demolish.

10.4 CONCLUSION

In the past decade, housing management of Dutch social landlords has changed, due to influences from the market and policy. Reductions in government regulation, financial support and social housing needs have brought Dutch housing associations in need of concepts to develop stock policy. In the past, asset management was mainly done on the basis of technical considerations, and major investments were mainly done as a part of externally (mostly government) driven renewal or refurbishment plans or on an *ad hoc* basis. Nowadays, there are several initiatives in the social rented sector to develop their own stock policy in a more structured way.

The Dutch regulations allow a lot of room for social landlords to determine their own asset management strategies. Although lower-income people are the main target group, social landlords can play a significant role in housing other income groups as well. The size of the sector alone (35% of the total Dutch housing stock) indicates that the social sector is far from residualised. Housing associations are allowed to sell their homes and are also allowed to develop owner-occupied housing. Individual housing grants facilitate social (and commercial) landlords to offer affordable housing of a fairly wide variety in quality. There are, however, some tendencies that will diminish the degree of freedom that social landlords have now. Some tax benefits have been dissolved and rent regulations are stricter. At this moment, it is unclear how strong the impact on asset management will be.

Market and financial factors have become more important in asset management of social landlords than they were in the past. Exacerbated by vacancy problems in the 1980s and early 1990s, lettability is high on the agenda of housing managers. However, it must be said that lettability is not a hot topic at present, because the rental housing market has become tenser in recent years, mainly due to a sharp rise of prices in the owner-occupied sector. Financial factors, of course, also play a part, but mainly in feasibility studies in the later stages of decision making and not in the early selection of policy options.

There are several initiatives among social landlords to develop a more structured top-down oriented asset management, in addition to the more traditional bottom up way of decision making. However, the results of this are modest. The goals and guidelines for the development of the mix of dwellings are, generally speaking, too general to have a serious impact on asset management decisions on estate level. Even if there are goals about the composition of the stock, they are not always 'translated' to goals on estate level. It seems that housing associations feel an urgency to develop a systematic policy, based on explicit guidelines, but hesitate to apply it on the estate level, because they appreciate a more unstructured, flexible way of decision making as well.

In general, we may conclude that, despite professionalisation tendencies, asset management is partly in its infancy. Furthermore, developments are slow. Research by Van den Broeke (1995) into the asset management practice of supposed fore-runners among housing associations in 1994 showed similar results, such as a weak relation between central objectives and actual strategies, the lack of a systematic approach in the form of a process model, and limited deliberation between alternative strategies.

These conclusions are in contrast with the results of a survey by Straub (1997, p. 35 and 82), in which 76% of the housing associations stated that they had developed a strategic stock policy (described in his questionnaire as 'all activities carried out by a landlord to ensure short-term and long-term congruency between the housing stock and the market demand as well as the landlord's objectives'). However, if stock policy of these landlords had actually been analysed, results would probably be different. This situation may change in the future, as in a recent survey by Primavera and van de Wijgert (2002), housing associations consider the development of strategic asset management to be one of the most important issues in their policy.

REFERENCES

Aedes (2001) Bedrijfstakinformatie 2000, Hilversum, Aedes.

Ansoff, H.I. (1984) Implanting strategic management, New Jersey, Prentice Hall International.

Boelhouwer, P. (ed.) (1997) Financing the social rented sector in Western Europe, Housing and Urban Policy Studies, 13, Delft, Delft University Press.

Boelhouwer, P. (1999) International comparison of social housing management in Western Europe, Netherlands Journal of Housing and the Built Environment, 14, pp. 225–240.

Broeke, R.A. van den (1995) Voorraadbeleid bij voorhoedecorporaties: model en praktijk, Delft, Delft University Press.

Broeke, R.A. van den (1998) Strategisch voorraadbeleid van woningcorporaties: informatievoorziening en instrumenten, Delft, Delft University Press.

CFV (2001) Verslag financieel toezicht woningcorporaties 2000, Naarden, Central Housing Fund.

CFV (2002) Het vermogen en de opgave van woningcorporaties: spanning tussen behoefte en middelen, Naarden, Central Housing Fund.

Flier, K. van der, and Gruis V. (2002) Portfolio analysis in social management, European Journal of Housing Policy, 2, pp. 183–202.

Flier, K. van der and Thomsen, A. (1998) The future of the Dutch housing stock; does the housing stock in the early post-war districts still have a future?, Proceedings ENHR conference 'Housing Futures: Renewal, Sustainability and Innovation', Cardiff.

Ginneken, C.C.P.M. van and Laag, B.G. van der (1992) Het proces en de organisatie van strategisch voorraadbeheer, in: Strategisch voorraadbeheer; 7 variaties op een thema, Lelystad, V-Plus, pp. 22–26.

Kam, G. de (2001) Uitkomsten onderzoek sociale huursector en grondmarkt, Aedes-Magazine, nos. 16–17, pp. 7–11.

Kempen, R. van and Priemus H. (2002) Revolution of social housing in the Netherlands: possible effects of new housing policies, Urban Studies 39, pp. 237–253.

Kersloot, J.M. (1999) (Des)investering- en huurprijsbeleid van verhuurders van woningen, Delft, Delft University Press.

Kullberg, J. (2002) Consumers' responses to choice-based letting mechanisms, Housing Studies, 17, pp. 549–579.

Leent, M.J.J. van and Vliet, J.M. van (1992) Om welke strategische keuzes draait het?, in: Strategisch voorraadbeheer; 7 variaties op een thema, Lelystad, V-Plus, pp. 4–11.

Marsman, G. and Smit, S. (1999) Feiten en achtergronden huurbeleid verhuurders 1999, Amsterdam, RIGO.

MVROM (2000a) Mensen-Wensen-Wonen: wonen in de 2le eeuw, The Hague, Ministry of VROM.

MVROM (2000b) Toezichtsverslag sociale huursector 1999, The Hague, Ministery of VROM.
MVROM (2001) Bouwprognoses 2001–2006, The Hague, Ministry of VROM.
MVROM (2002) Cijfers over wonen, The Hague, Ministery of VROM.
Nieboer, N., Heeger, H. and Haak, M. van der (2001) Strategisch voorraadbeleid: een ontwikkelingsmodel, Real Estate Magazine, no. 16, pp. 51–55.
Nieboer, N. and Gruis, V. (2001) A model for strategic stock management in social housing, Proceedings ENHR conference 'Housing and urban development in new Europe', Pułtusk, Poland.
Nieboer, N. and Gruis, V. (2002) What's strategic in asset management of Dutch landlords?, Proceedings ENHR conference 'Housing cultures; convergence and diversity', Vienna.
Otter, H.J. den and Heida, H.R. (2002) Primos prognose 2001, de toekomstige ontwikkeling van bevolking, huishoudens en woningbehoefte, The Hague, Ministery of VROM.
Primavera, A.F. and Wijgert, G.R. van de (2002) Onderzoek naar ICT bij woningcorporaties, Aedes-Magazine, no. 4, pp. 34–37.
Priemus, H. (2001) Social housing as a transitional tenure? Reflections on the Netherlands' new housing memorandum 2000–2010, Housing Studies, 16, pp. 243–256.
Remkes, J.W. (2002) Huurprijsbeleid 1 juli 2002 t/m 30 juni, Staatscourant, no. 21, pp. 28–31.
Straub, A. (1997) Strategisch voorraadbeleid en technisch beheer in de corporatiepraktijk, Delft, Delft University Press.
Straub, A. (2002) Strategic technical management of housing stock: lessons from Dutch housing associations, Building Research & Information, 30, pp. 372–381.
Sunikka, M. and Boon, C. (2002), Housing associations and sustainable management; environmental efforts in the Netherlands' social housing sector, Delft, Delft University Press.
Thomsen, A.F., Conijn, J.C. and Veerdonk, A. van de (2001) Vernieuwing Woningwaarderingsstelsel, Amsterdam, RIGO.
Vliet, J.M. van (1993) Integraal voorraadbeleid; naar een model voor het opstellen van beheerplannen op complexniveau door woningcorporaties, Delft, Delft University Press.

11 CONCLUSION

VINCENT GRUIS and NICO NIEBOER

11.1 INTRODUCTION

In many countries, housing management is changing in response to political and market developments. One of the current developments is the increasing interest in asset management in the social rented sector and the need for social landlords to develop a more strategic approach towards the management of their housing stock. In this book, the current state of play has been described in nine countries. The aim of these descriptions is, one the on hand, to fill the gap in international literature concerning the way in which social landlords manage their housing stock and, one the other hand, to enable an analysis of the relationship between landlords' practice and the context in which they operate. In this chapter we draw our comparative conclusions. We start with a comparative summary of the country monographs. Following the format that has been used in writing these chapters, we will subsequently describe the characteristics of the social landlords, their national context for asset management and their asset management practice. Then, we analyse the relationships between landlords' approach towards asset management and their national context in the light of the hypothetical model that has been presented in our introduction. We conclude with recommendations for policy and practice.

11.2 COMPARATIVE SUMMARY

11.2.1 Social Landlords

The share of social rented dwellings in the social housing stock varies between the investigated countries, as we can see in Table 11.1. The Netherlands has the highest share of social rented dwellings. Flanders (Belgium) and Australia are the countries in which the share of social rented dwellings is relatively low. In all countries the owner-occupied sector has the largest share of the total housing stock, although in Germany, the private rented stock is a close competitor.

Vincent Gruis and Nico Nieboer (eds.),
Asset Management in the Social Rented Sector, 209–225.

Table 11.1 Division of housing stock by tenure in various countries

	Social rented[1] (%)	Private rented (%)	Owner-occupied (%)
Australia (2002)	5.6	23.8	70.6
Austria (2000)	22	21	57
Belgium-Flanders (2001)	5.5	22	72.5
Denmark (1998)	20	18	58
England (2002)	19	10	71
France (1999)	20	25	55
Germany(1998)[2]	20	39	41
Latvia (2000)	30	9	61
The Netherlands (2001)	35	11	53

As far as ownership is concerned (local government, central government, company, private not-for-profit organisation, etc.) many countries have more than one type of landlord (see Table 11.2). France and Australia even have six categories of landlords. In some countries only one category of landlords exists or is dominating the social rented sector, such as the housing associations in Denmark, Flanders and the Netherlands. It should be noted, however, that within categories considerable differences exist between size, organisation, stock, etcetera.

11.2.2 National Context for Asset Management

General administrative regulations

As is common in the social housing sector, social landlords in all of the investigated landlords are subject to government regulations concerning their asset management. These regulations, however, vary from general guidelines or objectives for management, such as in the Netherlands, to detailed control by the (central) government, such as in Denmark. Furthermore, the instruments that the government uses to regulate the landlords' asset management vary. Generally, a mixture of instruments is used, but in some countries the emphasis lies on detailed legislation (e.g. Denmark, Belgium), in other countries on regulation by subsidisation programmes (e.g. France, Germany, Austria) or on supervision (e.g. the Netherlands). The nature of the regulations also varies

[1]In this table, the percentage of 'social rented' dwellings stands for the share of dwellings operated by social or public landlords. In the case of Latvia it means 'public sector' dwellings.
[2]Derived from tables 8.1 and 8.3. Percentage social rented sector is excluding social rented dwellings owned by private investors.

Table 11.2 Overview of social landlords in various countries[3]

Country	Social landlords	Average size (number of dwellings)	Percentage of social rented stock
Australia (2000)	Public (state) housing	45,371	79
	Community housing	24	5
	Crisis housing	5	1
	ATSI housing	31	5
	Government employee housing	1,000	6
	Defence housing	17,000	4
Austria (2000)	Municipalities	172	10
	Housing associations	2,632	13
Belgium-Flanders (2001)	Housing associations	1,309	100
Denmark (1998)	Housing associations	700	100
England (2002)	Local authorities	10,195	11
	Housing associations	842	8
France (2001)	Departmental "OPHLM"	8,000	7
	Municipal "OPHLM"	4,000	11
	Departmental "OPAC"	15,300	13
	Municipal "OPAC"	9,600	10
	Private housing companies	5,650	35
	Other social landlords	not available	24
Germany (1998)	Former social housing companies } Cooperative associations }	2,169	20
Latvia (2000)	State housing	not available	2
	Municipal housing	not available	98
The Netherlands (2001)	Housing associations	3,800	99
	Municipal housing companies	not available	1

according to the category of asset management activities which they concern (rent setting, allocations, sales, maintenance and renewal).

Rent setting

In almost all countries, rents are strongly regulated by the central government. In some countries, there is some room for landlords to differentiate their rents

[3]We have underlined the landlords whose asset management practice has been discussed in the country monographs.

(Belgium, the Netherlands, England, Germany, Austria), but this is considered as small.

Allocations

In most countries, the government sets income limits for households to be eligible for a social rented dwellings. Sometimes additional criteria are being imposed by the government to prioritise the allocation of housing to special target groups (disabled, retired, etc.; see e.g. Latvia). In some other countries allocation criteria are either absent (Denmark) or left up to the housing associations and/or local governments (e.g. the Netherlands).

Sales

National housing policies concerning sales are quite different. In some cases there is (or has been) an explicit policy to promote the sale of social rented dwellings, varying from 'soft' stimulation of landlords (France, the Netherlands) to outright privatisation policies, including the 'right to buy' (Latvia, England). In other countries the sale of social housing is highly restricted or even prohibited. In Austria, landlords can choose to give the tenants of new dwellings the option to buy in return for co-financing of the land and/or construction costs.

Maintenance and renewal

Central government regulation of landlords' maintenance policies is limited. Day-to-day maintenance is left up to the individual landlord, although in some cases the budgets for maintenance are subject to central regulation (e.g. the major repairs fund of the English local authorities). Regulation for renewals consists mostly of targeted subsidisation and investment programmes. In some countries there are no subsidisation and investment programmes for social housing renewal (Latvia, Belgium) or the programmes are not specifically targeted at social housing (e.g. the Urban Renewal Investment Fund in the Netherlands).

Financial support

In general, structural financial support for social landlords' asset management seems limited—subsidisation is (still) mostly targeted at new developments. There are, however, subsidisation programmes for urban renewal and housing improvements, which' size and conditions vary over time. In almost all countries, indirect financial support exists through some form of housing allowances, although the size and target group varies considerably.

In some countries, a financing system exists which takes into account the financial means of the individual landlords. In Denmark, the social landlords operate within a completely budget-neutral system in which it is impossible for an individual housing association to make a (temporary) profit. In Belgium, housing associations with a low rental income enjoy a slightly favourable loan repayment scheme and have been granted a special subsidy to meet the gap between rental income and expenditure. In the Netherlands, there is currently a debate about how to match the (substantial) differences between the tasks and means of individual housing associations. In England, (subsidised) stock transfers from local authorities to (new) housing associations have a substantial role in the refinancing of the local authority housing stock. For landlords in Austria, the lack of own means is a condition for being eligible for subsidisation of maintenance and renewal activities.

Supervision
Social landlords are either supervised:

- directly by central or local governments, for example: Australian State Governments, Latvian municipalities, English local authorities, The Dutch Housing Ministry;
- or indirectly by central, relatively independent organisations, for example: MILOS for the private housing companies in France, the Housing Corporation and Audit Commission for the English housing associations, the VHM for Flemish housing associations and the Central Housing Fund for the financial supervision of the Dutch housing associations.

Housing markets
In all of the investigated countries, there is (high) pressure on the social housing market. In some cases this pressure is due mostly to a general housing shortage (or shortage of decent, affordable dwellings), such as Australia, Belgium and Latvia. In other countries, the pressure is due mostly to the relatively good conditions in the social rented sector (compared to other sectors), combined with unrestricted access criteria, such as in Denmark and the Netherlands. However, within all of the investigated countries, there are regions, neighbourhoods and estates, with a threatened market-position in which social, technical and lettability problems occur. In some countries, such as France, parts of the social rented sector are also under threat by competition from other sectors (private rented and owner occupied dwellings).

11.2.3 Asset Management Practice

Rent setting
Due to the restrictive rent-regulations, the possibilities for social landlords to pursue their own rent policies is limited. Furthermore, housing allowances

can mitigate differences in the net rent level for the tenants. This means that an eventual rent decrease has a limited effect on the net costs for tenants and, consequently, a limited effect on demand as well. Thus, there are considerable limitations to the use f price as an instrument within asset management.

In most countries, rents are not market-oriented and in some they are not even related to the quality of the dwellings (Latvia, Flanders-Belgium). In the Netherlands and England, where there has been some leeway for housing associations to determine their own rents, social landlords have developed their own methods for assessment of the quality and/or market-position of their dwellings to support their rent policies. Recently, however, government regulations have become much stricter in both countries. In Germany, many landlords seem not to use their legal possibilities to set their rents in a more market-oriented way, due to political pressures and social traditions. Austrian housing associations are often forced to reduce their cost-based rents (by reducing the costs of finance) of the existing stock, due to the low rents of newly constructed subsidised rental dwellings, which are a result of the subsidisation system and which leads to the moving of tenants from the existing into the new stock.

Allocations

In countries in which government regulations allow it, landlords have developed or chosen their own allocation system and criteria, often in cooperation with the municipalities and other locally operating landlords. In the Netherlands, advert models (or choice-based lettings) have become most popular during the 1990s, replacing the more common system of waiting lists. But, while in England the advert model is also being introduced, it is already under discussion in the Netherlands and some Dutch housing associations have begun to experiment with lottery systems and the abolishment of allocation criteria. In other countries, allocations are strongly influenced by central regulations, although "some landlords make creative use of any available margin to pursuing their own allocation policies" (Flanders-Belgium).

Sales

The sale policy of social landlords reflects the national regulations to a great extent. In some countries, sales do not take place due to central restrictions and in countries where a right to buy has been established one can hardly speak of a sale policy of the landlord. However, in France, the Netherlands, Austria and Germany where sales are allowed and sometimes even stimulated (but not forced), social landlords do not sell of their dwellings in large quantities. In France, the private housing companies have only sold about 5,000 dwellings on average during recent years, while the government target was to sell about

30,000 dwellings. In the Netherlands sales have increased during the nineties, but the peak (in 1999) has remained below 1% of the housing stock. Since then, in parallel with the increasing pressure on the social rented sector, sales have dropped considerably, in contrast to the targets of the former State Secretary of Housing. On the one hand, these low numbers of sales can be explained by the lack of good market-opportunities, but on the other hand they can be explained from the social objectives and tradition of these landlords which lead them to be restrictive in their sale policy. In Austria and Germany, social landlords are reluctant to sell dwellings from the existing stock, among others because of the management problems associated with mixed-ownership. In Australia the State Housing Authorities are forced to sell of parts of their stock, despite the high pressure on the public housing sector, to cope with their financial shortages.

Maintenance and renewal

In general, social landlords seem to have an 'operational' approach towards their maintenance. Maintenance is either carried out on the basis of merely technical reasons and surveys or follows the complaints of costumers. In some cases, maintenance is restricted by the lack of financial resources (see e.g. Australia). Renewals seem to be determined mostly by subsidisation and investment programmes, rather than by initiatives from the social landlords themselves. But there are notable exceptions. In the Netherlands, housing associations have begun to implement market-oriented or client-oriented maintenance policies, aiming to increase the diversity of their housing stock and tenants' satisfaction. And in England, France and the Netherlands, some social landlords have taken on a leading role in the renewal of problematic neighbourhoods.

Decision-making process and methods

Most social landlords follow an incremental approach towards decision-making. Decisions are not based on systematic analyses and procedures, nor is there much use of rational frameworks or decision-support methods. In some cases, the decision-making process can best be described as 'muddling-through' (see chapter on Latvia). Market-orientation is limited among social landlords; they are mostly—or most of the times solely—focussed on fulfilling their social housing tasks. Partly due to the restrictions in which they operate, they do not actively seek for opportunities to make profit (which can be used to finance social housing tasks). Within their investment programmes, they focus on the problematic estates, rather than on a comprehensive analysis if the entire portfolio. Furthermore, there is little evidence of the occurrence of proactive behaviour; in general landlords seem to follow government investment programmes. In short, there is little occurrence of approaches towards asset management that can be described as 'strategic' as defined in the introduction of this book (being market-oriented, systematic, comprehensive and proactive). However, there are differences between countries and in particular between landlords within countries. A more strategic asset management is clearly being pursued in the social rented sector in the Netherlands, England, France and Australia. Some front-runners among landlords in these countries have already developed an approach towards asset management that can be typified as strategic. They have, for example, begun to:

- adapt decision-making frameworks (portfolio analyses) from the private sector to support the management of their entire housing stock;
- adopt market-oriented approaches towards their rent and maintenance policies;
- use sales and developments as structural instruments to generate financial resources for their social housing activities;
- take a leading, initiating role in neighbourhood renewal processes.

11.3 RELATIONS BETWEEN ASSET MANAGEMENT CONTEXT AND PRACTICE

In Table 11.3 we give an overview of key-characteristics of the landlords, their national context and their approaches towards asset management. Following our hypothetical model (Figure 1.5 in our introduction), we attempt to typify the relevant characteristics in general, internationally comparable terms such as 'restrictive', 'decentralised', 'market-oriented', etc. Furthermore, we reflect on the occurrence of strategic approaches towards asset management and the national factors which are stimulating such approaches. Judging from our analysis, in general the occurrence of 'strategic' asset management is limited, but we can see a tendency towards a more strategic asset management in some countries, while this is entirely absent in other countries. When we analyse these differences to identify relationships between the national context and asset management practice, we can not identify a clear, causal relationship, but we can identify a tendency towards strategic asset management among landlords who operate:

- with a certain degree of administrative and/or financial autonomy;
- a relatively large housing stock, with concerns about the market position of at least a part of their stock;
- with a professional organisation (see also Figure 11.1).

Table 11.3 Landlords' context, characteristics and asset management approaches

Country	Australia	Austria	Belgium-Flanders
Type of landlords (considered in this Table)	State housing authorities	Housing associations (HA) and municipalities (M)	Housing associations
Characteristics national context: (rental) housing system and market			
General administrative regulations concerning asset management	Unrestrictive, through general guidelines Commonwealth-State Housing Agreement	Relatively unrestrictive, through legal boundaries and Housing Promotion Scheme (HPS)	Restrictive through subsidies and prior authorisation
Rent regulations	Restrictive, based on household income	HA: restrictive; (M:) depending on subsidy	Restrictive, but with margins
Allocation regulations	Unrestrictive, decentralised	Unrestrictive, depending on HPS	Restrictive (central)
Sale regulations	Unrestrictive, decentralised	Unrestrictive; 'Option to Buy' for new dwellings	Unrestrictive
Maintenance and renewal regulations	Unrestrictive, decentralised	Unrestrictive; promoted though HPS; Maintenance component in HA rent	Unrestrictive
Financial support of asset management	No specific support for asset management, but general capital grants	Direct support through HPS; small indirect support	Direct support through VHM loans and project subsidies
Characteristics market (social housing need)	High pressure on social housing	General quantitative shortage, but oversupply in some sub-markets	General social housing shortage
Classification Kemeny	Dualist rental market	Unitary rental market	Dualist rental system
Characteristics landlords			
Objectives/ activities	Focus on low-income households	HA: fairly broad target group; M: mainly low-income households	Focused on low-income households
Size (average dwellings/ landlord)	Large (45,371)	HA: medium (2,632) M: small (178)	Relatively small (1,309)
Market position	Poor quality, but high pressure	Low (but improved) quality, low rents, local lettability problems	Good lettability due to high quality, low price and shortage
Organisation	Professional	Professional	Varying, larger landlords professional
Solvency	Poor	Generally healthy	Varying, poor in cities
Asset management approach landlords			
Market or task-oriented	Task-oriented	Tendency towards market-oriented, business-like	Task-oriented, small steps towards business-like
Systematic or unstructured	Tendency towards systematic	Unclear	Unstructured, small steps towards systematic.
Comprehensive or partial	Tendency towards comprehensive	Varying	Partial
Proactive or reactive	Tendency towards proactive	Varying	Reactive
Tendencies towards strategic asset management?	Yes	Some orientation on more strategic approaches	Very limited
(National) key-factors in the development towards strategic asset management	Financial independence and challenges and large size	Local concerns about market position	Initiatives from the supervisor (the Flemish Housing Company)

(continued)

Table 11.3 (Continued)

Country	Denmark	England	France
Type of landlords (considered in this Table)	Social housing associations	Housing associations (HA) and local authorities (LA)	Private housing companies
Characteristics national context: (rental) housing system and market			
General administrative regulations concerning asset management	Restrictive through subsidies and detailed legislation	Unrestrictive, mainly through approved programmes	Unrestrictive, mainly through subsidy programmes
Rent regulations	Restrictive	Restrictive	Restrictive, based on subsidy programme and location
Allocation regulations	Restrictive (decentralised)	Unrestrictive (decentralised)	Restrictive, based on income
Sale regulations	restrictive (not permitted)	Unrestrictive (L.A.: stimulating)	Unrestrictive (stimulating)
Maintenance and renewal regulations	Restrictive, National Building Fund (NBF)	Restrictive, through decency standard	Unrestrictive
Financial support of asset management	Direct support through NBF and subsidies and indirect support through housing benefits	Indirect (housing benefits) and direct (grants for renovations; but not for maintenance)	Direct through refurbishment grants and indirect through housing benefits
Characteristics market (social housing need)	Pressure on social housing due to low price and unrestrictive allocation criteria	General social housing shortage (varying per region)	General social housing shortage (particularly in large cities)
Classification Kemeny	Unitary rental market	Dualist rental system	Dualist rental system
Characteristics landlords			
Objectives/ activities	Broad target group	Focused on housing low-income households	Focused on low-income households
Size (average dwellings/ landlord)	Small (700)	Varying widely (HA: 842; LA: 10,195)	Relatively large (5,650)
Market position	Generally healthy, good lettability due to shortage	Generally healthy, some problem-estates	Generally healthy, with notable exceptions
Organisation	Democratic	Varying, larger landlords professional	Professional
Solvency	Financial surpluses are not possible	Varying, some landlords are financially threatened	Mostly poor
Asset management approach landlords			
Market or task-oriented	Task-oriented	Task-oriented, although increasingly business-like	Tendency towards market-oriented
Systematic or unstructured	Systematic (due to regulation)	Unstructured, with exceptions	Tendency towards systematic
Comprehensive or partial	Comprehensive	Partial, with exceptions	Partial, with exceptions
Proactive or reactive	Reactive (to regulation)	Generally reactive	Generally reactive
Tendencies towards strategic asset management?	None	Yes	Yes
(National) key-factors in the development towards strategic asset management	Virtually absent due to democratic tradition and restrictive regulations	Regulatory and financial challenges, local concerns about market position and LSVT	Regulatory requirements and local concerns about market position

(*continued*)

Table 11.3 (Continued)

Country	Germany	Latvia	The Netherlands
Type of landlords (considered in this Table)	Former housing companies (SHC) and cooperatives (CO)	Municipality	Housing associations
Characteristics national context: (rental) housing system and market			
General administrative regulations concerning asset management	Unrestrictive, mainly through subsidy programmes	Unrestrictive, mainly through legislation	Unrestrictive, mainly through general performance-'tasks'
Rent regulations	Unrestrictive, except for subsidised dwellings	Unrestrictive (decentralised)	Restrictive
Allocation regulations	Unrestrictive	Unrestrictive, decentralised	Unrestrictive, decentralised
Sale regulations	Unrestrictive	Stimulating through 'right to buy'	Unrestrictive (stimulating)
Maintenance and renewal regulations	Unrestrictive	Unrestrictive, decentralised	Unrestrictive
Financial support of asset management	Some direct support through specific funds; indirect through housing benefits	No central government support	Only indirect support through housing benefits and guarantee structure
Characteristics market (social housing need)	Generally balanced with large regional differences	High pressure on social housing	Relaxing market in 1990s, now increasing pressure
Classification Kemeny	Unitary rental system	Dualist rental system	Unitary rental market
Characteristics landlords			
Objectives/ activities	Focused on middle and low-incomes; some diversification out of social rented housing	Focus on low-income households	Broad: providing housing services with emphasis on social housing
Size (average dwellings/ landlord)	Medium on average (2,169) but varying widely	Not available	Medium-large on average (3,800) but varying widely
Market position	Varying regionally	No vacancies, but poor quality	Generally healthy, varying
Organisation	Professional, semi-public (SHC); self administered (CO)	Bureaucratic	Professional
Solvency	Varying	Poor	Generally healthy, varying
Asset management approach landlords			
Market or task-oriented	Some tendencies towards market-oriented	Task-oriented	Relatively market-oriented
Systematic or unstructured	Little use of (available) systematic procedural models	Unstructured	Unstructured, with exceptions
Comprehensive or partial	Generally partial	Partial	Partial, with exceptions
Proactive or reactive	Generally reactive	Reactive	Generally reactive
Tendencies towards strategic asset management?	Among some front-runners, but many landlords remain 'captive' of their tradition	None (muddling though)	Yes
(National) key-factors in the development of strategic asset management	Regulatory freedom and financial independence	Asset management is hampered by the communist tradition and the radical privatisation	Regulatory freedom, financial independence and local concerns about market position

Some landlords in the Netherlands, Germany, England, Australia and France can be seen as examples from the above. Strategic or business-like behaviour can also be stimulated more explicitly by government-regulations as is the case in France and England, where landlords are required to develop business or asset management plans. In many of the investigated countries (Australia, France, England) a trend towards a more strategic asset management stems from the challenge to combine substantial social housing task (poor stock, large need) with reduced financial support and (thus) does not coincide with a market-oriented, but with a task-oriented approach. In general, however, strategic asset management is in a development stage in all of these countries and substantial differences exist between landlords' practice within one country. In other examined countries (Belgium, Denmark and Latvia) strategic asset management seems non-existent. Landlords in Austria are said to be orienting themselves on a more professional, independent and strategic approach, but this has yet to be developed in practice.

In our introduction (section 1.5) we have also discussed the possible relationship between the type of housing system, as classified by Kemeny (1995) and the approach towards asset management. Our assumption was that a more strategic behaviour can be expected among landlords that operate within a unitary rental market. Our overview in Table 11.3, however, does not support the existence of such a clear, simplified relationship. The housing systems in the Netherlands, which can (arguably) be seen as the front-runner in the development of strategic asset management in the social rented sector, is indeed classified as unitary rental markets. However, other countries in which strategic asset management is an issue, such as England, France and Australia, are classified as dualist rental systems. Denmark, on the other hand, can be classified as a country with a unitary rental market, but strategic asset management is non-existent. And, although Germany can be seen as the 'role model' of a unitary rental market, many landlords have remained captive of their tradition and political ties. Brech (in this book) even 'accuses' them of waiting for state intervention to make them come into action, instead of taking forward their asset management by themselves. The Dutch case also illustrates that the tradition of social landlords has a large influence on their current practice. Although there is a clear tendency towards a more strategic asset management among housing associations in the Netherlands, many associations still have to transform themselves from task-oriented organisations into social entrepreneurs—this takes a lot of time.

In Figure 11.1 we have summarised the identified key-factors in a simplified version of our hypothetical model concerning the relationship between landlords' context, characteristics and approach towards asset management.

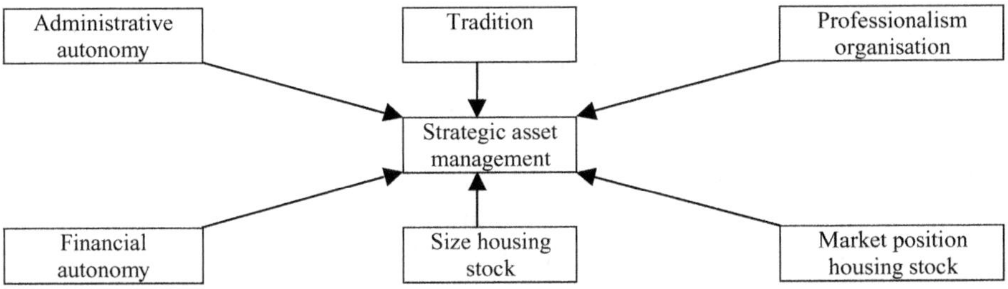

Figure 11.1 Identified key-factors in the stimulation of strategic asset management

11.4 CREATING THE RIGHT CONDITIONS FOR STRATEGIC ASSET MANAGEMENT

The knowledge of the relationship between the national context and asset management practice becomes particularly interesting if certain asset management approaches are recognized to have certain benefits and if these approaches could be promoted through national housing policies. In this final section we discuss the conditions under which strategic asset management can contribute to better practice in social housing management and the way in which strategic asset management can be stimulated through housing policy, using the knowledge of the key-factors in housing systems that stimulate their development. We conclude with a plea for a more structural international exchange of experiences between social landlords.

Conditions for 'good practice' in strategic asset management

In our introduction (section 1.3) we distinguished four general approaches within strategic asset management and mentioned their potential benefits for social landlords:

- market-orientation can contribute to a more efficient allocation of (financial) means;
- a systematic approach can contribute to a more transparent and rational asset management;
- a comprehensive approach can result in coherent policy without neglecting threats or opportunities in parts of the stock;
- a proactive approach can help to keep the housing stock in line with today's and tomorrow's needs.

These potential benefits support a plea for the stimulation of strategic approaches towards asset management in the social rented sector. However, as already mentioned in section 1.3, strategic asset management in the social rented sector is not without pitfalls. It's success depends highly on the conditions under which it is being developed and, if implemented under the wrong conditions, the strategic approaches could even 'backfire'. In the following paragraphs we reflect on these pitfalls and the conditions under which they could be prevented.

If market-orientation leads to market-conform behaviour, in a sense that social housing tasks are forgotten and business-efficiency and perhaps even profit-making will become the lead motive (see section 1.3), social landlords will loose their position as effective contributors to the fulfilment of social housing needs. In discussions in for example the Netherlands and England this is recognised as a serious threat within current housing systems. The stimulation of market-orientation in the social rented sector, therefore, must go hand in hand with a system in which social landlords are accountable for their performance as providers of decent, affordable social housing. Under such conditions, market-orientation could contribute to a more efficient use of the financial means and to tenants' satisfaction, by increasing client-oriented behaviour among social landlords.

The applicability of a purely systematic asset management in the social rented sector seems neither feasible nor desirable. It is not feasible since it is impossible to design a system for asset management that is able to cope with the full complexity of decisions that landlords face. Social landlords have to make choices based on criteria that are hard to compare and measure—criteria that can be placed under the general headings of social and financial return (see e.g. Kemp, 1995; compare, for example, with the discussion of the limitations of strategic planning by Mintzberg, 1995). However, social landlords can surely benefit from a more systematic decision making framework, just as long as these systematic procedures are not applied too rigidly. For example, decision-making frameworks derived from portfolio-analyses in the commercial sector, which are now used by some social landlords in France and the Netherlands, can help to reach more rational and transparent decisions (see also van der Flier and Gruis, 2002).

Comprehensiveness—in a sense of analysing of and setting goals for the development of the *whole* stock in stead of focusing on problematic estates—could lead to inefficient use of the organisation's resources; why should a landlord invest time in the analysis of estates which have no (apparent) problems? To prevent this potential inefficiency, comprehensiveness should be accompanied by (automated) management information or monitoring systems which can generate a 'quick scan' of the housing stock according to a few key-variables (vacancies, turnover, tenants' complaints etc.). Then, housing managers at a strategic level can identify (arising) problems and are able to compare the different estates within their portfolio with each other. Furthermore, there should be organisational procedures that enables and stimulates employees at an operational level to transfer their assessments of developments in estates to the management. Of course, these recommendations apply in particular to larger social landlords.

The choice between proactive and reactive behaviour is a difficult balance to be struck. On the one hand, it is generally acknowledged that 'to prevent' is better than 'to cure'. But, on the other hand, if a landlord really acts (and invests) before potential problems have become reality, this may also be considered to be an inefficient use of resources. Therefore, proactive behaviour must go hand in hand with a cost-benefit analysis of following an active versus a passive strategy.

Stimulation of strategic asset management through housing policy

If implemented under the right conditions, all of the above approaches within strategic asset management can lead to a more effective and efficient asset management in the social rented sector. And, reading the conclusions of the country monographs, many authors seem to recognise the potential benefits of strategic asset management, even authors from countries in which it is entirely absent in today's practice. However, implementation or stimulation of strategic asset management in these countries will not be simple due to barriers in the existing housing systems. Judging from the experiences in other countries and the key-factors that have been identified in section 11.3, strategic asset management could be stimulated in various ways. If central governments would choose to do so, key-aspects of their policy would have to be to stimulate the independence and professionalism of social landlords.

The stimulation of independence should not primarily take place by reducing financial support to social landlords. Rather, governments should give landlords the means and freedom to develop towards more financially and administratively independent organisations (see e.g. the Netherlands, France, Germany). This also implies that social landlords could be allowed to make some profit, at least on a part of their activities, so they can generate the financial means to cross-subsidise their primary activities concerning the provision of decent and affordable homes and to become solvent enough to operate more independently from financial government support.

From the cases of Germany and the Netherlands, where the regulatory context is relatively favourable for the development of strategic asset management, we can learn that landlords need a lot of time to change from bureaucratic, political, task-oriented organisations into professional social entrepreneurs. Therefore, next to setting the right regulatory context, governments could choose to stimulate professionalism and strategic behaviour among social landlords in a more direct way. This could be done among others by implementing procedures for performance measurement or by explicitly requiring social landlords to develop asset management plans (e.g. England, Flanders-Belgium, France). However, the governments should take care that such a top-down stimulation of professionalism by prescribing more systematic procedures for performance measurement and decision making will not become in contradiction with the stimulation of administrative independence, as has been stated to be the case with regard to the 'new public management reforms' concerning English local authorities which eventually lead to an increasing degree of central compliance and regulation (see Brown et al., 2002). A more indirect way to stimulate professionalism and the need for strategic asset management is the promotion of mergers, since asset managers within a larger landlord have more need to take a strategic view and, at the same time, larger landlords have more financial means to maintain a professional organisation. Nevertheless, the promotion of mergers has to coincide with awareness of its consequences for, among others, local housing management and monopolisation within the social rented sector and, thus, may not always be a favourable option.

This book indicates that 'best practices' are scarce because of the early stage of development of asset management in the social rented sector. As some of the interesting examples we mention:

- the more comprehensive and systematic approaches towards portfolio analysis that are applied by some Australian, Dutch, English and French landlords;
- the use of sale, rent and maintenance policies to reach a more efficient and effective allocation of means (see e.g. examples from the Netherlands);
- the implementation of methods for performance measurement and benchmarking in support of management decisions (see e.g. England and Belgium).

It is exactly in the current stage of development that landlords can benefit the most from exchange of experiences, particularly between landlords in countries in which there is a trend towards a more strategic approach. Therefore, a more structural international communication between landlords in these countries seems highly recommendable. Hopefully, this book will prove to be a good starting point for this exchange of knowledge.

REFERENCES

Brown, T., Richardson, J., and Yates, N. (2002) Whose asset management? Paper presented at the ENHR conference "Housing cultures; convergence and diversity", Vienna.

Flier, K. van der, and Gruis, V. (2002) Portfolio analysis in social management, European Journal of Housing Policy, 2, pp. 183–202.

Kemeny, J. (1995) From public housing to the social market; rental policy strategies in comparative perspective, London, Routledge.

Kemp, P.A. (1995), Researching housing management performance, Urban Studies, 32, pp. 779–790.

Mintzberg, H. (1994) The rise and fall of strategic planning, New York, The Free Press.

APPENDIX A: FORMAT FOR COUNTRY MONOGRAPHS

APPENDIX A: FORMAT FOR COUNTRY MONOGRAPHS

1. Social landlords in country X
   The first paragraph contains a short description of landlords in country X, following these questions:
   - What kind of social landlords exist in country X (local authorities, housing associations etc.)?
   - How large is their stock (total number of dwellings, average number of dwellings per landlord, share in total housing stock)?
   - What is their function on the housing market (do they only house minimum-income households or do they have a broader target group and active in the area of project development of owner-occupied and high-rent housing, are they highly independent or strongly regulated and supported by the government)?

2. National context for asset management
   A short description of the national context, following these sub-paragraphs:
   2.1. Regulations
   - What are government regulations for the landlords' asset management concerning: general administration (objectives and tasks landlords), rent setting, allocations, sales, maintenance and renewal, and tenant involvement?

   2.2. Supervision
   - How are the landlords supervised (by whom and on what basis)?

   2.3. Financial support
   - What financial support exists for the landlords' asset management (exploitation grants, subsidies for renewals, government loans, government guarantees, housing grants)?
   - Are there large differences in the financial position of social landlords; are there initiatives to 'match' social tasks and financial means of landlords, e.g. by mergers, money-transfers, or stock-transfers?

Vincent Gruis and Nico Nieboer (eds.),
Asset Management in the Social Rented Sector, 227–229.

2.4. Housing market
- What are quantitative and qualitative shortages on the (social) housing market?

3. Asset management practice

This is the core-section, where a general description is given of social landlords' practice, following these sub-paragraphs:

3.1. Rent setting
- How do landlords determine the (desired) rent level and yearly rent increase for their dwellings?
- Do landlords (have possibilities to) raise the rents in relation to tenants income?

3.2. Allocations
- What system do landlords use to allocate their dwellings (e.g. waiting list, advertise model) and what kind of criteria do they use in their allocations (e.g. income, household size, age, waiting period, etc.)?
- What legal protections do tenants have? Are there possibilities to evict them when their income becomes to high in relation to the rent?

3.3. Sales
- Do landlords sell dwellings and, if so, how do they determine which dwellings to sell and under which conditions are the dwellings sold (e.g. with discounts or sale-back guarantee)?
- What are the motives for sale: market, money or idealism (owner-occupation as a merit good)?

3.4. Maintenance and renewal
- How do landlords determine the maintenance level for their dwellings (e.g. only on the basis of stock condition surveys and technical need, or also related to preferences of the target group for the dwellings)?
- When do landlords decide to carry out more substantial renewals (e.g. when subsidies are available, when considerable social problems exist, when substantial vacancies arise (reactive), when the market position is expected to deteriorate (pro-active))?

3.5. Decision making process: considerations, methods and strategies
- What activities do landlords undertake to support their asset management plan (e.g. mission statement, market analysis, portfolio analysis, SWOT-analysis, strategy formulation, strategy evaluation, monitoring of dwellings' performance)?
- Are these activities part of a systematic process (e.g. sequential steps) or are the activities performed on an 'ad hoc' basis?

- Is the performance of the housing stock evaluated on a regular (periodical) basis or are analysis only performed when problematic estates occur?
- How are stakeholders, and tenants in particular, involved in decision making?
- Structuring the decision making process
- Do landlords use some kind of a rational decision making framework (e.g. by linking outcomes of the performance of dwellings to pre-defined management strategies)?
- What kind of (automated) decision support systems are used to inform asset management?

4. Conclusion

APPENDIX B: LIST OF AUTHORS AND THEIR AFFILIATIONS

Eva Bauer, Österreichischer Verband Gemeinnütziger Bauvereinigungen, Austria

Frédéric Bougrain, Centre Scientifique et Technique du Bâtiment, France

Joachim Brech, Planning and Research in Wohnbund-Network, Germany

Lars A. Engberg, Danish Building and Urban Research, Denmark

Joe Flood, University of Canberra, Australia

Vincent Gruis, Delft University of Technology, Faculty of Architecture, Department of Real Estate & Housing, The Netherlands

Nico Nieboer, Delft University of Technology, OTB Research Institute for Housing, Urban and Mobility Studies, The Netherlands

Andrew Thomas, HACAS Chapman Hendy, United Kingdom

Sasha Tsenkova, University of Calgary, Faculty of Environmental Design, Canada

Sien Winters, Catholic University of Leuven, Higher Institute for Labour Studies, Belgium

Vincent Gruis and Nico Nieboer (eds.),
Asset Management in the Social Rented Sector, 231.

GPSR Compliance

The European Union's (EU) General Product Safety Regulation (GPSR) is a set of rules that requires consumer products to be safe and our obligations to ensure this.

If you have any concerns about our products, you can contact us on ProductSafety@springernature.com

In case Publisher is established outside the EU, the EU authorized representative is:

Springer Nature Customer Service Center GmbH
Europaplatz 3
69115 Heidelberg, Germany

Batch number: 10823959

Printed by Printforce, the Netherlands